ONE HUNDRED INNOVATIONS FOR DEVELOPMENT

Edited by Gillis Een and Sten Joste

Intermediate Technology Publications *in association with* the International Inventors Awards 1988

Published by ITDG Publishing
The Schumacher Centre for Technology and Development
Bourton Hall, Bourton-on-Dunsmore, Rugby, Warwickshire, CV23 9QZ UK
www.itdgpublishing.org.uk

First published in 1988
Print on demand since 2004

ISBN 1 85339 095 X

A catalogue record for this book is available from the British Library

ITDG Publishing is the publishing arm of the Intermediate Technology Development Group.
Our mission is to build the skills and capacity of people in developing countries through the dissemination
of information in all forms, enabling them to improve the quality of their lives and that of future generations.

The primary aim of the International Inventors Awards (IIA) is to advance development by giving
recognition to technical innovations which help to solve sociological and economic problems,
especially in the Third World.

Printed in Great Britain by Lightning Source, Milton Keynes

TABLE OF CONTENTS

PREFACE

When the King of Sweden handed over the International Inventors Awards – IIA – to the prize winners in the City Hall of Stockholm on 13 June 1986 it was a climax achieved after 10 years of planning and work. It was made possible by the patient and faithful support of the Salén Foundation, the Swedish International Development Agency – SIDA – and by the work devoted to the project over all these years by Professor Carl-Göran Hedén and many others.

But the IIA was conceived as a process that should not end with the prize-giving ceremony. At an early stage it was recognized that the real value of IIA for the developing countries does not lie in the awards given to a few of the many nominees. We received so many valuable nominations from all parts of the world covering a wide range of subjects that it would be a waste to let them stay idle in our files.

When the nominations began to arrive in great numbers in the summer and autumn of 1985 we realized that we would receive enough material for a book about the hundred best of them. In the autumn of 1986 we promptly obtained support from SIDA for the editing of the book. Thus we are now able to complete the IIA process and supply the very important feed back of innovations, inventions and good ideas to the people who need them in their work for technical, economical and social development in the Third World.

In all our undertakings we have met with a lot of encouragement from people of all professions and from all corners of the world. A majority of them expressed the hope that IIA would continue into another cycle of nominations, awards and feed back to the developing countries. We, who have been directly involved in the whole IIA process, are grateful for this opportunity to show what has been achieved and we sincerely hope that there will be more International Inventors Awards in years to come.

Gillis Een
Sten Joste
Stockholm 1987

IIA 1986

In 1795, almost 200 years ago, Napoleon instituted a prize for the best solution to the problem of providing his soldiers with easily portable, endurable packages of food. This stimulated a flood of innovative activities, which, through the prize winner François Appert, led to the invention of the canning process. We have here one of the first instances in modern times of how the setting of awards may contribute to important innovations. This is precisely what the International Inventors Awards are aiming at, although in a more peaceful context and with special emphasis on development problems in the Third World.

As a result of the organizational work we received within the stipulated time and from proper nominators a total of two hundred and sixty-one proposals from fifty different countries all over the world. Sixty per cent of these proposals originated from inventors in Third World countries and forty per cent from people in industrialized countries. In both these groups the majority of the most valuable proposals emanated from innovators working in research and development institutions, which specialized in appropriate technologies for Third World countries.

These two hundred and sixty-one proposals deal with problems concerning almost all aspects of life in the developing world. Together they give a highly instructive survey of what the international community of innovators with special insight regarding basic needs in these countries look upon as their most important and imminent problems.

It is perhaps no surprise that the biggest single group of proposals concern water: how to dig wells, how to store water, how to pump it, how to clean it, how to economize in its use, how to distribute it and how to achieve new and more efficient irrigation systems.

In the energy field we have received proposals regarding wind energy, solar energy, wave energy, biomass and geo-thermal energy. We also have proposals dealing with hydro-electric energy in small units for local consumption. But the perhaps most important energy problem in the Third World, occupying the minds of many innovators, consists in achieving increased efficiency in the use of firewood when cooking food.

Even if there was no special area for agriculture, numerous proposals, particularly within the industry group, take up problems connected with the production, drying, storing, conservation and distribution of food and also the use of agricultural by-products. There is of course particular attention to tropical crops, especially rice, sugar and palm oil.

Another important problem area richly represented among the received proposals concerns simple methods for house construction using local building materials. A special problem which is reflected in several proposals concerns the construction of simple, hygienic and water-saving toilets in rural communities.

In the forestry group we have received comparatively few proposals, most of them dealing with methods for the felling and planting of trees, the drying of timber and with the introduction of new species of trees.

As already pointed out, this comprehensive inventory of innovations gives a host of instructive lessons to all who are occupied with promoting economic and social development in the Third World. In order to take advantage of this unique asset, the IIA secretariat is preparing this book, presenting the one.hundred most important of the nominated inventions. The cost of writing this book will be borne by the Swedish International Development Agency (SIDA).

In its evaluation work the Prize Committee of IIA and its four subcommittees soon realized that the most important proposals were never limited only to technical details. In most cases the innovator has to struggle with a multitude of interconnected technical, social and economic problems. This is particularly the case in the rural areas of developing countries, where the majority of the population lives. Here the challenge to the innovator is especially difficult as such areas are as a rule poor, sparsely populated, with little infrastructure and almost no regular market structure. On the technical side of such socio-technical innovation packages the real problem is not to produce sophisticated novelties at the frontier of Western technology but rather to develop already known techniques in order to adapt them to local conditions. This means the finding of technical solutions which are simple, cheap, robust, easy to handle, to maintain and to repair, involving a minimum of accident risks and which can be produced, at least to a substantial degree, in regional workshops operating close to the areas of demand.

On the social side the innovator has to deal with the problem of integrating the new technical constructions into the existing social system with its firmly rooted habits and traditions. This poses intricate problems of social acceptance and diffusion, the handling of which often requires sustained efforts of information and education. In many cases also the organizing of new forms of co-operative labour and ownership is necessary as well as new forms of distribution and of financing.

Our study of the 261 proposals taught us that in order to bring forth innovations which fulfill the requirements prescribed by the statutes of IIA, the innovator has to be both researcher, inventor, entrepreneur and social field worker in one person. These are indeed very demanding and very rare qualifications. In the opinion of the Prize Committee of IIA they are amply fulfilled by all the nine prize winners.

In concluding it would be proper to underline what is often forgotten, namely that in order to develop innovations that are viable in the Third World you need as least as much time, money and professional qualifications as are required when developing innovations suitable for the industrialized countries. But in the latter case, with well organized markets and comparatively rich consumers, the chances of high returns are much greater. As a consequence, people who devote themselves to innovations for the poor areas of the Third World run particularly high risks and deserve to be given extra support and stimulus. It is the hope of the IIA organization that the prizes will constitute a good step to this effect.

— — ◊ — —

A slightly modified version of the talk that Dr Erland Waldenström, PhD, gave on the occasion of the IIA 1986 in the City Hall of Stockholm on 13 June.

THE PAST AND THE FUTURE OF IIA

The aims of IIA – the International Inventors Awards

The primary aim of IIA is to advance development by giving recognition to technical innovations which help to solve sociological and economic problems, especially in the Third World.

The target areas of IIA

Here are some brief and selective notes on the four target areas of IIA: forestry, energy, water and industry.

Forestry and energy

Agriculture in the wider sense of the word of course also includes forestry. Its most important role in the Third World is that it serves as a source of energy. Of the 1800 million m³ wood which are cut yearly in these countries some 80 per cent are used as fuel. According to a recent FAO report half of the world's population – some 2,000 million people – rely on wood and charcoal for cooking, space heating and even lighting.

This, coupled with the often ruthless exploitation of the tropical forests by industrialized countries, has led to the forest area in the Third World being reduced by between 8 and 20 million hectares per year or, as an average, 0.6 per cent of the total forest area.

Add to this depletion the fact that the industrial exploiters of the tropical forests usually select the kinds of timber which are most in demand on the world market, which means that the forests are seriously impoverished ecologically.

Only some 1.1 million hectares of forests are replanted in the tropics. This afforestation, obviously insufficient per se, is furthermore done unevenly: half of it in Latin America and, generally speaking, not where it is most needed, i.e. where there are human settlements. Furthermore, the ecological system of the natural forests can never be replaced by afforestation.

By introducing more efficient charcoal-conversion kilns or charcoal-burning stoves it is technically possible to double or even triple the resource availability from wooded land. However, this efficiency approach must be combined with long-term solutions implying the development of agro-forestry systems in which wood production complements food yields in a non-competitive way. This is a complex problem, offering many opportunities for innovational thinking.

In more general terms, research into new and renewable sources of energy, and conservation of existing resources, must continue and be intensified, since energy is one of the most essential resources for the intensification of agriculture and food production, and for industry.

Energy, in fact, ranks in importance with the classical factors of production – land, labour and capital. Energy innovation requires an understanding of the role of technology in development as well as of societal energy needs.

4

Water

About 97 per cent of the world stock of water is constituted by the salty waters of the oceans. Of the remaining fresh water, 75 per cent is frozen and not readily available for use; 96 per cent of what is not frozen is underground and not easily accessible, and it is only a tiny fraction of the freshwater in rivers and lakes which, together with atmospheric precipitation, is readily available for our use.

Furthermore, there are several constraints on the water which we may use:
- the hydrological cycle implies that dry areas remain dry and wet areas continue to have abundant water;
- catchment areas are often destroyed by man or from natural causes;
- population growth increases the demand on useable water beyond available resources;
- where sufficient quantities may be at hand, the quality of the water may be unsatisfactory.

These constraints and others make water a scarce resource even though it covers such a vast part of the planet. The use of this resource has to be carefully planned. Any innovation in planning the handling of water resources must aim at removing or reducing the constraints just mentioned and making use of the water most effectively to contribute to development without endangering its quantity and quality.

What is primarily needed is not so much new mechanical gadgets and devices for the pumping or transport of water but rather such innovations that foster the proper management of the resources, the conservation and re-use of water, through an integrated societal approach.

Industry

In IIA parlance, industry presumes low investment and capital costs and is based on methods of utilizing local resources without exhausting them. Optimization and preservation are the key words.

Examples of the activities of this target area are:
- the food chain extending from the production of basic food materials right through methods of preparing meals but with a stress on the need for improved methods to preserve and upgrade basic food materials;
- methods to prevent waste and spoilage of food are very relevant; special cases in this context are industrial processes making it possible to utilize indigenous and little-known food crops;
- housing; especially such innovations are relevant which utilize locally available building and raw materials;
- clothing and transport are other examples from this target area.

Amongst the criteria for innovations in the industrial field low investment and capital costs as well as utilization of local raw materials have already been mentioned.

Another criterion is that employment opportunities should be located as near as possible to where the majority of people live – that is, mostly in the rural areas – so that large numbers of people can avail themselves of new work opportunities without having to move to towns.

Finally, production methods must be simple so that large numbers of people can use them without relying on sophisticated skills, organization, or materials.

New techniques – winners and losers

New techniques and new knowledge like all development create winners and losers. On this theme Carl Tham, Director General of Sida, has made some perceptive remarks:

'...Perhaps the most important explanation of the mismatch of needs and technology on the village level (is) the fact that poor people cannot afford to put anything at stake such as is often required when something new is to be introduced. They cannot afford to lose. Indeed smallholders and landless people are, and must be, extremely risk averse and careful in their adaptation of new means and methods of earning their living.

This risk aversion among the rural poor has caused frustration among many well-meaning aid workers. They come equipped with fine ideas and tools which could improve life for the potential beneficiaries. Yet their intentions have to submit to very careful evaluation by the community they work in before, maybe, their ideas are tested...

Aid to promote innovations at the grass roots level is a very difficult undertaking which takes a long time, hard work and skilled and enthusiastic staff. In view of the difficulties of bringing about change through innovation in this setting, the fact that the (IIA) award winners have managed to create inventions that have been successful at the village level increases our admiration for their work even further...'

As a consequence of the built-in scepticism towards innovations, those who are seeking to find solutions to the social and economic problems causing poverty in the developing countries usually receive little material or moral support from their governments. There is, as a rule, no training of inventors. Nor is there any public relations apparatus to promote their products. Foreign goods are favoured. The technological co-operation between developing countries is still insignificant.

IIA: encouragement to inventors – governments – industries – aid agencies, retailers of new ideas

Against this background, awards of the IIA type may help to encourage and stimulate inventors and to draw public attention – and, in particular, the attention of the authorities – to their work. Policy makers must be made aware of the fact that many socio-economic problems can be solved given sufficient motivation and resources, and that the kind of climate must be created within which the innovatory process can flourish.

There is thus a need for a comprehensive system of encouragement :
- to the inventors themselves;
- to the governments, industries and aid agencies which have the means to facilitate their work; and
- to the 'retailers' of new ideas (entrepreneurs, extension workers, credit organizations and the like) without whom inventions cannot move out from the laboratories into the market place, in ways which enable them to be taken up and used by the intended beneficiaries.

IIA document: 'Innovations for development'

With this in view, a considerable amount of brain storming took place at various levels shortly after the idea of IIA was conceived. A major symposium resulted in the publication of a basic IIA document, *Innovations for Development* . This book chronicles many of the early thought processes, and the symposium itself served a springboard from which to launch other aspects of the programme. The publication has served as a guide not only to the IIA but also to other ventures contemplated in the field of intermediate technology.

Operational framework of IIA

Following this, an organizational framework was created which devolved the executive authority to the IIA Secretariat and to target area committees. They identified the most worthwhile innovations and channelled them to an awards committee, comprising individuals with long experience of developing countries and also of high international reputation, for final selections.

It is, of course, impossible for an awards system – however well endowed – to reward directly all those involved in the long chain of events between the successful development of an innovation and its ultimate use by large numbers of poor people.

However, IIA has managed to create a large network of voluntary assistance, with more than 4,000 organizations and individuals involved all around the globe – collecting, sifting and forwarding nominations. The United Nations family – particularly the UNDP – has been most helpful; so have many academies, research institutes, inventors' organizations, patent organizations, etc.

The evaluation was modelled on that of the Nobel Foundation, a process greatly facilitated by the fact that the Chairman of IIA's Board was formerly for many years the President of the Nobel Foundation.

Possible modifications of target areas

The present subject areas of IIA – water, industry, forestry, and energy (WIFE) – have already been mentioned. There is obviously a certain degree of interdependence and overlap between them. This is as it should be and need not be avoided through a different terminology. It should rather be seen as a challenge to understand better the linkages between needs and resource utilization.

Food production and processing
More important is the question whether *food production and processing* might not be more strongly emphasized than now, given its central importance to the well-being and development of the large majority of rural people living in developing countries.

Transport
Should the number of target areas be increased, the London-based Intermediate Technology Development Group feels that *transport* should be seriously considered:

'...Transport – or the lack of it – is one of the greatest constraints suffered by people living in the rural and indeed the urban areas of developing countries.

Unlike other sectors, the planning and development of transport systems is still governed largely by western and inappropriate concepts. Expensive road systems are built with little thought for the real requirements of the majority of the people, or for the kind of vehicles which will use them. Their planning is governed by the need for the government agencies to have access to rural populations – not by an assessment of the needs of the people themselves.

Yet, even a cursory visit to any rural area, or to the slums of large cities, reveals the need to improve methods of cheap and simple transport in the form of headloading, wheeled vehicles, or animal carts as well as the need for cheap and easily maintained and constructed roads and tracks along which such forms of transport can travel.

Developments in this field involve not only appropriate designs of roads and vehicles, but also appropriate methods of manufacture and distribution. Since the burden of transport falls mostly upon women, improvements to the systems would have far ranging social and economic advantages, Without appropriate methods of transport, rural people have neither the means nor the motivation to increase their productivity or to benefit from the increase...'

The IDRC feels that the possibility of extending the target areas to cover such areas as public management, economics and rural development should perhaps also be explored.

Development of the award system

Before considering further possible changes in the IIA award system, some brief notes on its development up to date may be of interest.

In order to stimulate both technical and social creativity for the benefit of the poorest people in the Third World, the Swedish Inventors Association decided, in 1976, to launch a form of prospective awards, first suggested at a meeting of the International Federation of Institutes for Advanced Studies (IFIAS).

Funds were granted by the Swedish Salén Foundation, and it was decided that the awards should be conferred in 1986, ten years after the first announcement of the target areas had been made. The award distribution in 1986 would coincide with the centenary of the Swedish Inventors Association, the oldest of its kind in the world and also one of the largest.

Information on the aims of the awards was released at simultaneous press conferences held in Stockholm at the Association's Headquarters and in Geneva at the United Nations World Intellectual Property Organization (WIPO). It was also widely distributed to several hundred organizations and individuals all over the world together with a request for suggestions of innovations suitable to be rewarded.

As a result, 361 proposals were received from fifty countries on different continents, in time for the award distribution by H. M. the King of Sweden in June 1986. Of these proposals 80 per cent originated from inventors in Third World countries and 40 per cent from people in industrialized countries. In both these groups the majority of the most valuable proposals came from inventors working at research institutions specialized in developing technologies appropriate for the Third World countries.

The biggest single group of proposals concerned *water*: how to dig wells, how to store water, how to pump it, how to clean it, how to economize in its use, how to distribute it and how to achieve new and more efficient irrigation systems.

In the *energy* field strikingly many inventors work on increasing efficiency in the use of fire wood for food cooking. However, suggestions were also received regarding wind energy, solar energy, wave energy, biomass and geothermic energy, as well as proposals dealing with hydro-electric energy in small units for local consumption.

Many proposals, particularly within the industry area, referred to problems connected with the production, drying, conservation and distribution of *food*, and also the use of agricultural by-products. Of course special attention was given to tropical foods, especially rice, sugar and palm oil.

Another important problem area richly represented among the received proposals concerned simple methods for *house construction* using local building materials. A special problem which was reflected in several proposals concerned the construction of simple, hygienic and water-saving toilets in rural communities.

In the *forestry* group comparatively few proposals were received, most of them dealing with methods for felling and planting of trees, drying of timber and with the introduction of new species of trees.

This comprehensive inventory of inventions gave a host of instructive lessons on the promotion of economic and social development in the Third World. In order to take advantage of this unique asset, the IIA Secretariat has prepared a book, presenting the 100 most important of the nominated inventions. The cost of writing this book was borne by SIDA.

According to the IIA statutes it is expected that a prize-winning innovation should lead relatively quickly to positive economic and sociological results. In fact, one lesson to be drawn from the study of the proposals made for the 1986 selections is that the innovator, in order to fulfill the IIA requirements, has to be researcher, inventor, entrepreneur and social field worker in one person.

These are indeed very demanding requirements, but they were amply met by all the prize-winners.

Prize-winners

Those were, in 1986, the following innovators:

The water prize

Dr Peter Morgan (born in England) was awarded a shared prize for having been the driving force behind three innovations: a handpump, a bucket pump and a ventilated pit latrine, all produced at the Blair Research Laboratory in Harare, Zimbabwe, where he has been working for many years.

These products are simple, can be manufactured at low cost and maintained on site; furthermore they suit the social patterns of both rural and urban populations. They are widely used for water supply and sanitation in Zimbabwe and neighbouring countries.

Mr Vilas Salunke, India, was awarded a shared prize for his system for optimal water conservation in poor areas of rural India prone to water shortages.

Thanks to his in-depth knowledge of local natural conditions he has been able to develop a socio-technical system, which has given substance to the concept of fair water distribution, based on size of family and not on ownership of land. The potential for expansion of this system is considerable.

The industry prize

This prize was awarded to Dr Amir Kahn for his development of the Asian Axial Flow Thresher, a light rice thresher of simple design and easy to manufacture and service. It makes it possible to change over to a more intensive form of agriculture, producing a grain yield which is up to 10 per cent higher than with conventional methods, without upsetting the normal patterns of society. It is now made by more than 600 small manufacturers and more than 50 000 threshers now exist in Asia alone.

The forestry prize

This award was given to Michael Benge, James Brewbaker (both Americans) and Mark Hutton (an Australian).

In their work with the species *Leucaena leucocephala* they have done important pioneer work in developing ways to stretch still remaining forestry resources, and further develop and enlarge them. They have stimulated the rapid production of firewood, construction material etc. and reduced the pressure on other remaining forests. But *Leucaena* will also make possible the enlargement of the forestry area generally, by providing a fast-growing multi-purpose crop.

The energy prize

This prize was divided in two equal parts. Half of the award went to Mr Peter Fraenkel for his construction of a low-maintenance, long-life wind-powered water pump to ensure water supply and irrigation in remote areas of developing countries. He has managed to combine traditional wind energy technology with modern engineering and manufacturing principles to solve an urgent problem: to base rural water supply in developing countries on local renewable water resources.

The other half of this award went to Mr. Armstrong Evans and Gerald Pope for their construction of an electronic load controller for small hydro-power stations.

One of the advantages of this innovation is to bring down costs for regulating the power output from turbines to match load, which has always been one of the main problems.

As a result, total energy costs are decreased and the number of sites available for economic exploitation is increased.

Mr. Evans and Pope have managed very successfully to integrate modern power electronics with simple small-scale hydro-power technology adapted to the needs of developing countries in a way that makes cheap, reliable and effective load control possible in small turbines.

Award amount

For each area the award amounted to Swedish Crowns 250,000 (at 1987's rate of exchange over US $ 40,000).

Since the public recognition of the innovator and his work is the essential factor rather than the size of the award, it would seem that the amounts need not be increased, at least not within the next few years.

IIA – an award or a process?

The IIA awards should be considered not as the per se primary aim of the IIA activities but as the visible manifestations of a development process. This process implies a mapping out of technical needs in various parts of the world of different thematic kinds, bringing those needs to the attention of policy-making authorities, evaluating and awarding suitable technologies and – last but certainly not least – following up the translation of the technologies awarded into practical solutions integrated in various social systems.

Future financing

Given the successful completion of the first round of the IIA exercise, it is highly desirable to keep the momentum and establish a firm financial basis for future awards.

So far the Swedish Salén Foundation and SIDA have granted the necessary funds. Henceforth, also foreign agencies should be invited to join in this venture, either by contributing to the award system as a whole or by sponsoring one or other of the individual awards. Here are some indications of the preliminary reactions from a potential donor:

'...As we told Mr. Joste, we have indeed been impressed with both the agenda and the procedures that have been established by you and your colleagues to enhance the incentives and visibility for technical innovations adapted to the poorer regions of the world. Mr. Joste's explanations and the supporting materials he has presented to us have been informative and helpful. We are sympathetic to the IIA's objectives and supportive of the steps that have been taken to implement the programme.

From our conversations we understand that the financing for the awards and the implementation of the program hopefully will be covered by certain funding sources. The Rockefeller Foundation is prepared to give sympathetic consideration to project activities which support or enhance the impact of the awards program and we advised Mr. Joste that we would welcome the opportunity to review your planning for such activities. I might tell you in advance that activities based in the developing countries are generally given higher priority by the Foundation.

Our general response therefore is a positive one and we look forward to hearing further from you...'

The Department of External Affairs, Canada, and UN agencies second the positive appreciation of IIA's activities and feel '...that this (IIA) is a very worthwhile program and that it should be continued if at all possible...'

Periodicity

All this to ensure a prize-distribution say every three years, a periodicity which would permit a complete process cycle and allow enough time for the necessary collection and sifting of proposals and information.

Broadening the base of IIA

In this connection it would be appropriate to make certain administrative adjustments. Whereas the executive functions should remain in Swedish hands, the advice of non-Swedish personalities of world-wide reputation in their respective fields should be sought; the modalities for such consultations to be further considered. By thus broadening the base, IIA may more easily attract foreign funding, at the same time giving the awards a wider international status.

Follow-up support

A question to be considered is to which extent IIA might give follow-up support to worthwhile innovations in ways other than by the award of prizes. This is a difficult and time-consuming process, requiring professional on-the-spot counselling and assistance to innovators, to help them get their products on to the market.

In addition to the assistance to entrepreneurs, the activities of IIA should also include *research*, *education* and *information*, whereas development projects should be left to the already existing NGOs, so as to avoid duplication of efforts. This should not preclude IIA from seeking ways and means of stimulating other agencies to take up and promote innovations themselves to the point where they actually result in benefits to people.

Inventions, of themselves, are of little importance unless they are taken up by, and lead to real benefits for, the intended users. The ITDG, e.g., therefore feels that

'...a prime consideration when making awards must be proof of the fact that the innovator has made a proper assessment of the risks, and has taken or proposed realistic measures to minimize them for all concerned.

Therefore a special award might possibly be given for the best example of the marketing of an innovation, where the problems and prejudices attached to its use have been successfully overcome. Such an award could be made to an extension worker, community developer, entrepreneur or others–indeed to the inventor himself... The emphasis would be less on the technical or social implications of the project, but on the recognition of the problems to be overcome, and the ways in which they were tackled in the process of bringing the technology to the market...'

- - o - -

An abbreviated version of the report that Dr. Sture Linnér submitted to SIDA after his evaluation of the IIA project.

THE HUNDRED BEST INVENTIONS

The purpose of this book is to present the hundred best inventions among the many nominations received for the International Inventors Awards of 1986. We present them in two groups of fifty each.

Before publication we have approached the majority of the nominators related to the hundred best inventions and have received very positive replies from most of them. This is remarkable as the addressees are scattered all over the world and many of them work in very remote places.

Perhaps we should also point out that some of the inventions mentioned do not meet all the criteria set for the awards, but are nevertheless of such interest that the editor feels that they warrant a place in this book. Naturally the choice of the 'hundred best' sets a rather arbitrary limit.

In the first group of fifty inventions we have placed the best of those which have been described to us in detail and which we judge as being of interest to a large number of readers and potential users. We have given them a page each, where they are described in text and illustration. A uniform style of language and editing is used and each description contains the name and address of the person to contact in case readers are interested in using the invention in their own country.

In the second group we have placed fifty inventions which in most cases have not been fully described to us. Here we have limited the presentation to a short descriptive title and the name of the person to contact. It may be difficult to reach him/her, but write to us and we shall do our best to help you.

In each group the inventions are listed according to the four target areas water, industry, forestry and energy. Within each area the prize-winning innovation(s) is presented first, followed by those which received an honourable award. The order of those inventions which follow thereafter does not reflect any ranking from our side.

We publish this book as a follow-up of the award itself but also as a part of an evaluation of the whole IIA-project. We do this in anticipation of a second set of International Inventors Awards in the near future. For this reason we would very much appreciate every kind of reader reaction or comment. We have every reason to believe that the IIA has contributed to the development of new appropriate technologies and to the broadcasting of information about what is available worldwide today. We look forward to hearing from you.

'Water for all' is the slogan for the United Nations' *Drinking-water Supply and Sanitation Decade, 1981-1990* . The 73 nominations to the IIA *target area water* together reflect all aspects of the UN programme in this field. Many of the proposed inventions were pump constructions (20 per cent), often with a long record of the technical, economic and social aspects of successful field trials. Many methods of water purification were presented. An important part was played by complete systems for water distribution and management. Of particular interest are a number of proposals aiming at reducing or eliminating the need for water in deling with night soil and similar sanitary problems, thus reducing the pollution load on scarce water resources.

Bertil Hawerman

In our final request for nominations to the IIA *target area industry* we suggested a concentration on the food chain, housing, clothing and transport. After a first rough screening, which eliminated all the complicated and resource-demanding suggestions, we were left with 43 proposals for serious consideration. Nine of those referred to agricultural tools and methods. Fourteen referred to food processing, preservation or storage. Fourteen referred to methods or elements for building houses, including the use of local building material. Only two referred to clothing and one to transport. Finally there were three proposals entirely outside the areas mentioned here. We have recorded that in many cases the innovations had created work in local workshops and small industries, which was exactly what we had hoped for.

Gillis Een

We were rather disappointed by the fact that we received less than 30 proposals for the IIA *target area forestry* . This is after all an area of great importance not only for the production of timber and pulp but also for soil conditioning, erosion protection, crop shelter, fuel for cooking and more. Among the best proposals we found a number of improved designs of e.g. saw teeth or protective sleeves, but also new systems for forest management and the use of new species of forest trees.

Mårten Bendz

Most of the nominations to the IIA *target area energy* consisted of improvements on well-known technologies. They covered all forms of energy and usually implied better utilization and better distribution of existing sources of energy.

Lars Kristoferson

THE NAIGAON EXPERIMENT

PRIZE WINNER

Mr Vilas Balwant Salunke
Pune
INDIA

A socio-technical system for water conservation and water management in a dryland farming area in India. Water is collected, infiltrated and finally distributed according to the needs of each farming family.

This is a system of water management in an area with uncertain or irregular rainfall. The strategy is to maximize the agricultural production in good rainfall years and to minimize the losses when rainfall is inadequate. This is achieved by conservation of runoff from the watersheds of a village by the construction of tanks, reservoirs, barrages and percolation tanks for infiltration at upper levels, and by harnessing the water directly from these water storages by digging open wells below, fitted with electrically operated pumps.

The water is shared in the villages according to the number of members in each family and *not* in proportion to the land holdings. The irrigation schemes are undertaken for groups of dryland farmers and *not* for individuals. Each person gets enough water to irrigate 0.20 hectare of land over the whole year. If a family owns more land than this unit, the rest of the land has to rely on natural rainfall. The scheme also prohibits crops requiring a lot of water.

After three years of operation in Naigaon village the production of cereals was twenty times larger than in previous years of drought.

The scheme also includes the education and training of young unemployed people to serve as agricultural consultants.

There is a wealth of documentation material including a video cassette produced by the Tata Institute of Social Science.

There are no patents or rights to cover the innovation.

Contact: Mr V. B. Salunke, Gram Gaurav Pratisthan, 670 Hadapsar Industrial Estate, P.O.Box 1202, Pune 411 013, INDIA. Telephone 70158

THE VENTILATED IMPROVED PIT LATRINE

PRIZE WINNER

Dr Peter Morgan
Blair Research Lab.
ZIMBABWE

A ventilated pit latrine which is almost odourless and which meets high sanitary standards. It requires no water and is thus particularly suitable for arid areas.

The ventilated pit latrine consists of a pit dug in the ground and lined with cement, plaster or bricks and a concrete cover slab with two holes i.e. one for squatting and one for the ventilation pipe. A roofed building covers the structure. Air is drawn up the ventilation pipe and is replaced by fresh air passing down the squatting hole. The ventilation pipe is fitted with a corrosion resistant fly screen – usually made of fibreglass or stainless steel – with the double purpose of preventing flies from entering the latrine and acting as a death trap for flies trying to leave towards the light through the ventilation pipe.

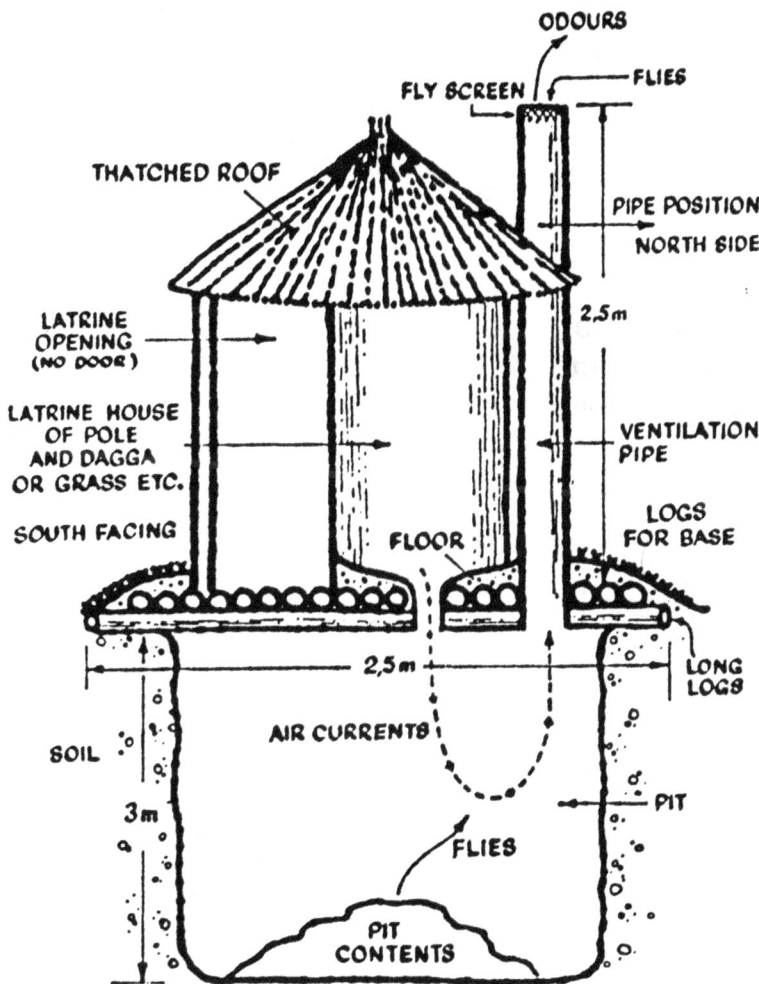

This latrine does not require any water and is therefore particularly suitable for geographical areas where water is scarce.

Between 1975 and 1985 over 100,000 latrines of this design have been built in Zimbabwe alone and the rate of construction is increasing.

Contact: Dr Peter Morgan, Blair Research Lab., Harare, ZIMBABWE.

THE BLAIR HANDPUMP

PRIZE WINNER

Dr Peter Morgan
Blair Research Lab.
ZIMBABWE

A handpump consisting of two pipes, one movable inside the other. The inner pipe has a check valve in the bottom and a water discharge spout in the upper part. The spout also acts as a handle.

The Blair Handpump uses simple principles, and is little more than two pipes, one fitted inside the other, with a non-return valve at the lower end of each. Mass produced models were designed for groups of up to ten families, i.e. for between 50 and 100 people. The pump delivers 15 to 20 litres per minute from a depth of not more than 12 metres. The pump has no levers and no water seals and the handle doubles as water spout. The outer pipe works like a cylinder and the inner pipe like a piston, push-rod and water spout all in one.

A complete pump can be taken out, inspected, all the valves replaced and reassembled within ten minutes. There are no nuts or bolts, and tools are provided with each pump.

Like all Blair designs the pump can be made by hand or purchased commercially. Details of hand built pumps are on the school curriculum in Zimbabwe and instructional literature is available on the home construction of both heavy duty and light duty models.

The pump is cheap, the lightest model costing about USD 90.

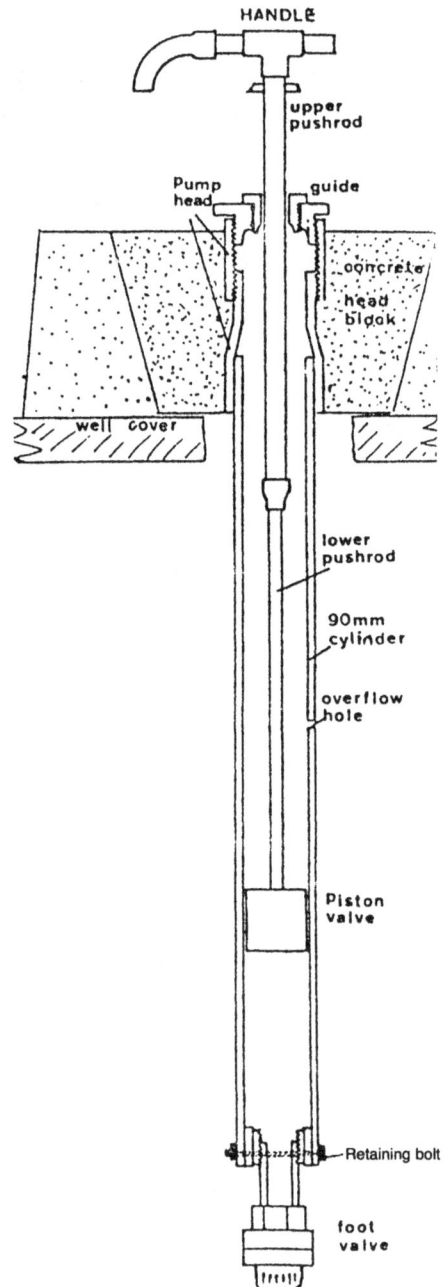

Contact: Dr Peter Morgan, Blair Research Lab., Harare, ZIMBABWE.

THE BLAIR BUCKET PUMP

PRIZE WINNER

Dr Peter Morgan
Blair Research Lab.
ZIMBABWE

A water lifting device consisting of a well tube and a rather closely fitting narrow cylindrical bucket provided with a check valve in the bottom.

Of all the water raising machines the bucket and windlass is the most common throughout the world. This simple concept has been incorporated in a modern design called the Bucket Pump. It went on trial in Zimbabwe in 1982 and was mass produced in 1984. About 1000 Bucket Pumps are now in operation and the interest is growing.

The pump consists of a narrow cylindrical bucket and a simple poppet type of valve which provides a fast filling rate. A special water discharge device consists of a funnel on which the tubular bucket is placed and the bottom valve is opened automatically. This protects the pump and the water from contamination and helps discharging into the waiting family water vessel.

The bucket carries 5 litres of water and can lift it up to 20 metres.

The pump is easy to understand, use and maintain in an ordinary farming village. The Blair institute provides detailed technical information, which is provided with a multitude of illustrations.

Contact: Dr Peter Morgan, Blair Research Lab., Harare, ZIMBABWE.

THE FLOATING WATER PURIFICATION PLANT

Mrs Zhenwan Xu
Gansu Province
CHINA (PRC)

A new type of water purification plant. It is mounted on a barge and constitutes a relatively cheap and self-contained unit for the production of potable municipal water from river or lake water i.e. the water it is floating in.

The Floating Water Plant (FWP) is designed for countries with rivers and lakes as the only sources of municipal water. FWP is a self-contained unit mounted on a barge. The draught has been reduced to between 1.0 and 2.5 metres and it can thus be moored near the shore.

The intake of raw water from 0.8 metres below the surface is by means of a siphon. The sedimentation is enhanced with the aid of inclined tubes. The filter is provided with a multi-media bed containing e.g. grains of rubber, anthracite and sand.

The inventor claims the following advantages:
- FWP can be constructed in a shipyard and the capital costs are reduced to 50% of the conventional.
- FWP can be moved if better needed elsewhere.
- FWP does not occupy valuable farmland.
- Pumps are to a large degree replaced by siphons.
- FWP is flexible and can adapt to variations in raw water quality.

16 plants have been built so far. The throughput has been between 500 and 40,000 cubic metres per day of potable water. A plant for 100,000 cubic metres per day has been designed.

Patents are pending in China.

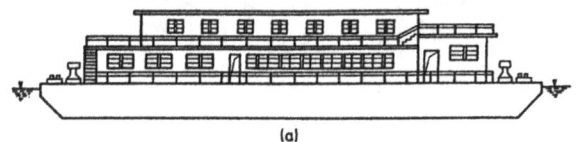

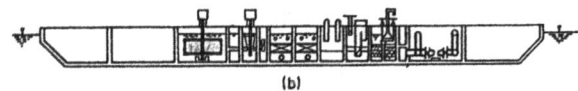

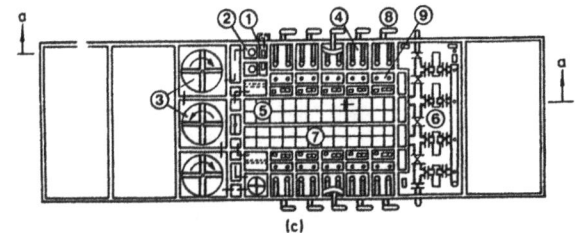

(a)

(b)

(c)

① siphon intake
② mixing chamber
③ coagulating basin
④ sedimentation basin
⑤ distributing channel
⑥ pumping station
⑦ filter
⑧ sludge discharge siphon
⑨ water collecting through discharge siphon

Contact: Mrs Zhenwan Xu, Northwest Municipal Engineering Design Institute, 177 Ding Xi road, Lanzhou, Gansu Province, CHINA.(PRC) Telephone: Lanzhou 24713.

THE AUTOMATIC THROTTLE HOSE

Prof. Daniel L. Vischer
Mr. P.J.Peter
SWITZERLAND

An outlet regulating device for small irrigation systems, which delivers a near-constant flow independent of the water-head in the reservoir

In irrigation management, as well as storm water engineering, there is a need for a near-constant discharge of water from a reservoir independent of fluctuations in the upstream water level. This innovation is a regulating device which is sufficiently exact and at the same time simple to construct and maintain. It is cheap and can therefore be afforded also in small irrigation systems and in the peripheral parts of larger systems.

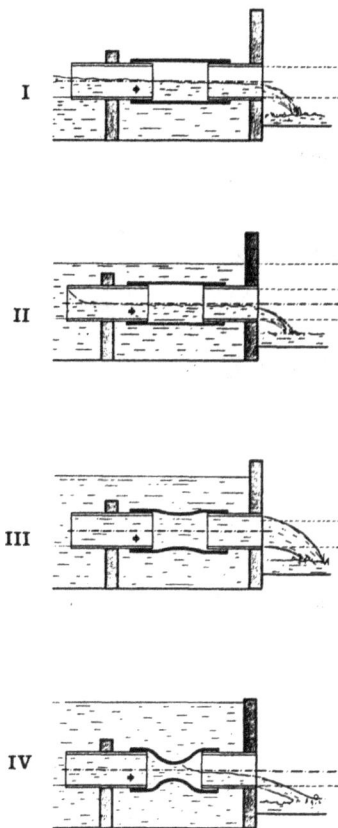

The regulating device is placed near the bottom inside the reservoir from which the water is discharged. It consists of a fairly long pipe followed by a short flexible hose. The pipe provides a pressure drop when water is flowing through. The hose is deformed and compressed in proportion to the water head in the reservoir and to the pressure drop in the pipe. The end effect is a throttling action which keeps the flow near-constant. There are often vibrations in the device but this helps to prevents silt deposits and clogging. By modifying certain components it is possible to extend the range of the regulating device. It can deliver 100 litres per second at the farm and 1 cubic metre or more further upstream in the irrigation system. The throttle hose has been tested in tropical irrigation in TOGO for more than two years with good results.

Patents granted in one country and pending in two.

Contact: Dr H. Brombach, UFT Umwelt- und Fluid-Technik GmbH, D-6990 Bad Mergentheim, GERMANY (BRD).

SWEDHEART PUMP SYSTEM

Mr. Eric Bruce
Karlstad
SWEDEN

A water pumping system consisting of two simple displacement pumps. One pump creates a pulsating pressure in an operating fluid and the other pump discharges water to the surface.

The semiautomatic Swedheart Pump System consists of two separate displacement pumps – one for the operating medium and one for the water.

The operating medium is ordinary water of good sanitary quality contained in a closed system. A peristaltic pump of special design, called "ORIGO", creates a pulsating pressure but only a small reciprocating flow in the operating fluid. The ORIGO pump can be driven by man or beast and is placed away from the well head thus avoiding contamination of the water from urine etc.

The water pump proper is called "HEART". It consists of a closed chamber containing a pulsating rubber bladder which is connected to the operating liquid. The chamber is provided with two check valves. The one in the bottom opens and lets in water when the bladder shrinks. The one in the top opens and lets water into the discharge pipe when the bladder expands.

The HEART pump does not require any particular anchorage in the well and can be moved up or down in order to adjust to the water level. It can easily be taken out from the well for inspection and maintenance. It is robust and can stand rough handling.

The design capacity of the HEART pump is 0.6 - 50 cubic metres per hour against a head of up to 80 metres. Patents are pending in several countries.

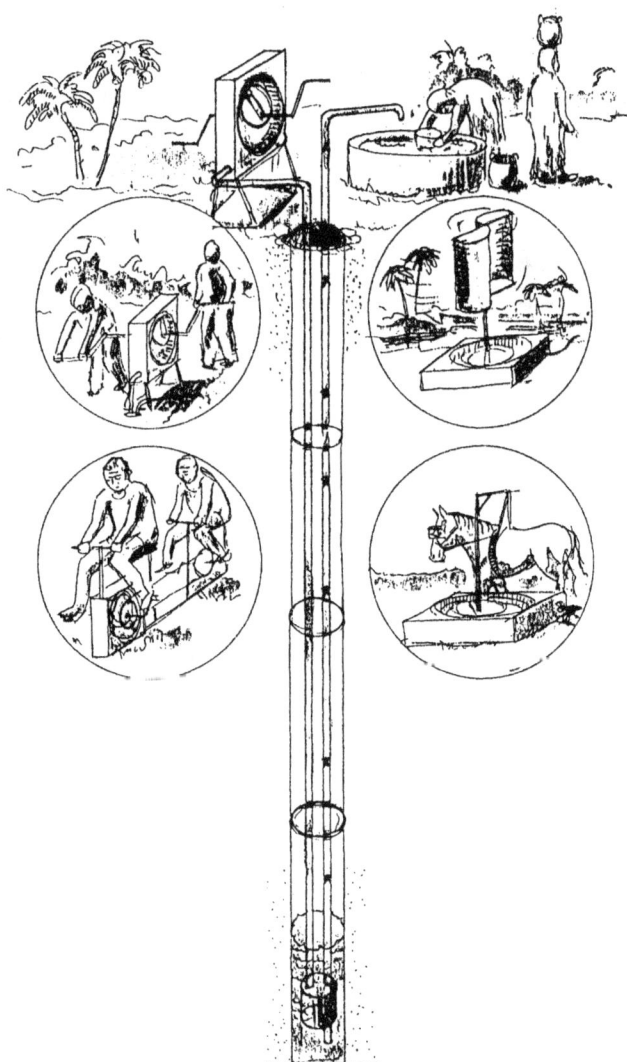

Contact: Dr. Bertil Hawerman, CTI, Skeppargatan 3, S-11452 Stockholm, SWEDEN. Telephone: +46-8-679000. Telex: 15331 CTI S

THE TREADLE PUMP

Mr Gunnar Barnes
RDRS
BANGLADESH

A very simple and inexpensive pump operated by leg-power. It can be manufactured and maintained locally. It is used for irrigation on very small farms.

The main components of this extremely simple pump are: a bamboo tube well; two identical cylinder pump heads; a frame and two treadles of bamboo; ropes of nylon. The pump is powered by a walking action – hence the name. It is manned by one or two adults or two to four children, who often enjoy it as play. The throughput is 160 litres per minute with a water table of 3 metres and 90 l/m at 6 m. It can be manufactured in local rural workshops and maintained by the farmer himself without problems. It is expected to last for five to ten years before replacement is required.

This pump is most suitable for very small landowners. The pump is operated by the farmer himself and his family. It would be too expensive to buy the required labour. The pump makes it possible to irrigate a small crop during the dry season. The cash obtained will in most cases pay for the pump in one season.

More than 50,000 treadle pumps are now in operation in Bangladesh alone.

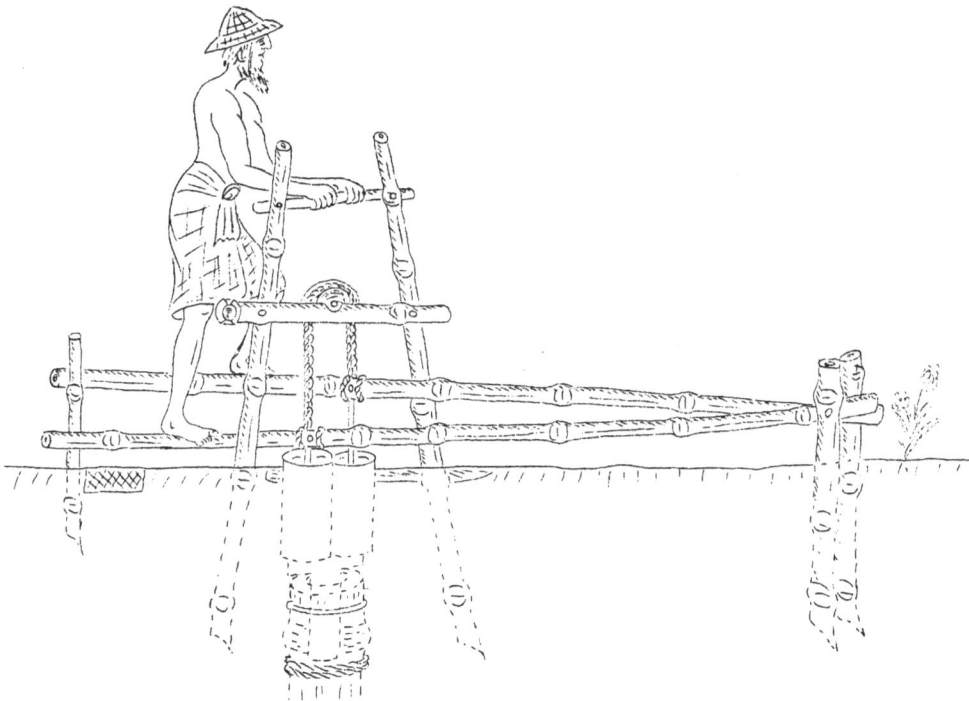

Contact: Mr Thorben T. Peterson, Director, Rangpur-Dinajpur Rehabilitation Service (RDRS), GPO Box 618, Ramna, Dhaka 2, BANGLADESH.

PORTABLE TOILET

Dr Abul Basher M. Shahalam
Yarmouk University
JORDAN

A portable toilet system for temporary habitats in places with a dry climate.

This invention consists of a portable toilet system with the following features: a sit-on toilet bowl with a water seal/gas lock at the bottom; a replaceable cylindrical filter cartridge which can be lifted out, cleaned and refilled with local sand; a waste water tank; a water recirculation and flushing unit provided with a hand pump.

Night soil is flushed down through the gas lock by means of recirculated waste water via the hand pump. The filter cartridge separates solids from the liquids. The latter is collected in the waste water tank in the bottom of the toilet. Organic material is degraded by anaerobic microorganisms. When the filter tends to clog it is taken out, cleaned and provided with fresh sand.

The main parts of the toilet are made of light weight plastic such as PVC in order to make it portable. The toilet is designed for temporary habitats such as places of pilgrimage, camps of nomads or refugees, emergency operations, urban slums etc.

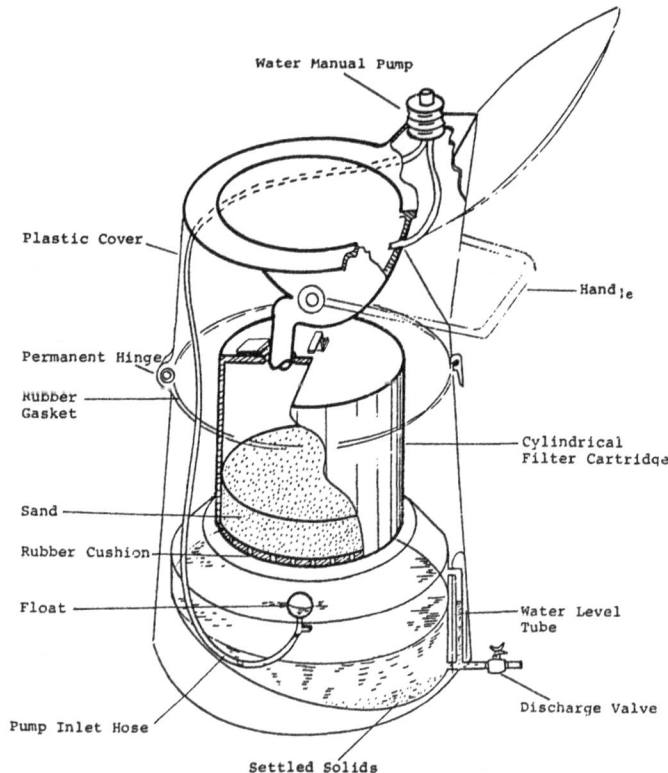

Contact: Dr Abul Basher M. Shahalam, 914 Eleventh Street, New Brighton, Pennsylvania 15066, USA

THE AXIAL FLOW THRESHER

PRIZE WINNER

Dr Amir U. Khan
International Rice Research
Institute (IRRI)
PHILIPPINES

A small portable thresher and winnower for freshly harvested paddy rice. It is manufactured locally, is cheap to maintain, gives a high yield and operates also in tropical rain.

Several sizes and versions of this thresher have been developed at the International Rice Research Institute. The smallest one weighs 100 kilogrammes and is portable. It can be carried out to the ricefields and can be placed on soft ground. It has a throughput of 400 kg per hour of freshly harvested paddy rice.

The machine is a combination of thresher and winnower. It separates the grain from straw and chaff. The separation is carried out by a combination of sieves and counter current air streams.

The thresher works even when it is raining. It has a low power consumption and is easy to maintain. It relieves the villagers of back-breaking work in the middle of the harvest when there is a shortage of labour. Threshing losses are as low as 1 % which should be compared with 3 - 6 % with conventional methods.

An estimated 70,000 axial flow threshers have been produced in 8 Asian countries by about 600 small rural manufacturers.

Contact: The Head of the Agricultural Engineering dept., IRRI, P.O.Box 933, Manila, PHILIPPINES. Tel: +63-2-7420717

FISH SMOKING TECHNOLOGY

HONOURARY AWARD WINNER

Mr B. Kagan
FAO
GHANA

A simple but effective fish smoker, which can be built and operated by village people. It produces smoked fish of high quality that keeps well in a tropical climate.

The following description of the Chorkor Smoker has been taken from a brochure in English produced by UNICEF. "The fish is placed on wire net nailed to rectangular wooden frames. Eight to fifteen frames filled with fresh fish are placed on top of each other on a rectangular oven built of clay bricks or cement blocks. The oven is divided into two equal compartments in which fire-wood is burned. The trays fit to each other and to the smooth upper surface of the oven, forming a chimney through which the hot smoke passes. The fish is thus hot-smoked with a minimum of heat loss. Two or four times during the process the order of the trays is reverted. At the same time the trays are turned 180 degrees on the oven and the individual fishes are likewise turned. In this way all the fish receives the same treatment."

The Chorkor Smoker gives a higher fish keeping quality and less waste. The consumption of firewood is reduced. The smoker can be built and operated by village people. Its use has spread to Benin, Guinea and Togo and information has been sent to South America.

Contact: Mr J.A.C. Davies, Assistant Director-General, Regional Representative for Africa, FAO, P.O.Box 1628, Accra, GHANA.

WAREHOUSE DRYER

HONOURARY AWARD WINNERS

Dr Yong Woon Jeon
Miss Leonides S. Halos
Dr Clarence W. Bockhop
IRRI,PHILIPPINES

An integrated system for drying a variety of agricultural products and food such as grain, tubers, fish and meat.

This innovation is a method for drying food using non-conventional sources of energy. The complete system consists of three main components i.e.:

- A Centre-Tube Furnace to produce drying heat from agricultural waste material such as rice chaff or straw.

- A Vortex Wind Machine to discharge the humid air from the drier.

- A Slotted Vertical Tray Set to carry the material to be dried. This includes a special weaving device for making tray bottoms out of coconut fibres.

All parts are stationary and there is a minimum of wear and maintenance. The driers are operated by villagers who find them easy to use. The heating costs are low as are the post harvest losses of e.g. rice due to rot and moulds. The driers are constructed from mainly local materials.

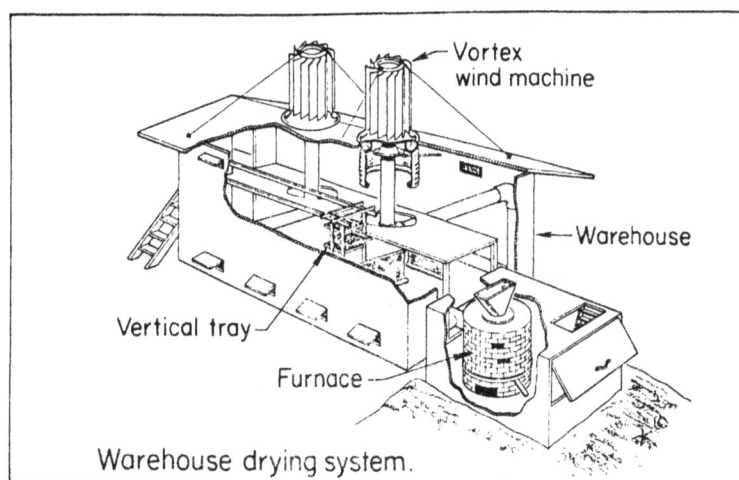

Warehouse drying system.

Contact: The Head of the Agricultural Engineering dept., IRRI, P.O.Box 933, Manila, PHILIPPINES. Tel: +63-2-7420717

ANIMAL DRAWN MULTIPURPOSE FARM IMPLEMENT

Mr Jean Nolle
Limours
FRANCE

A flexible multipurpose system of farm implements drawn by a pair of animals.

This innovation consists of a modular and flexible system for providing all the implements that are required on a farm in a developing country. The investment is minimal. No engine is required as the implements are drawn by a pair of oxen or horses.

The system can provide a number of functions. The most important are the following:

- Ploughing
- Harrowing
- Transport, including a tank for liquids
- Sowing, e.g. cotton and cereals
- Fertilising
- Harvesting, e.g. ground-nuts or forage
- Spraying biocides

The work on this system has been going on for a considerable length of time. Three generations of farm implements have been developed and there are light as well as heavy versions.

Cliché J. NOLLE.

K-NOL équipée en sous-soleuse.

Contact: Mr Jean Nolle, 8 rue des cendrière, F-91470 Limours, FRANCE. Tel: 4581516

SALT TOLERANT CROPS

Dr Mahmoud Abdel Kawy Zahran
Mansoura University
EGYPT

A study of salt tolerant plants with the aim of developing new crops for the production of paper pulp or animal forage.

This project is an investigation into salt tolerant (halophytic) vegetation in Egypt and other countries with the aim of developing new crops for subtropical areas with saline soils and/or with saline irrigation water.

The main part of the work has been a study of species of the genus Juncus (rushes). It has been shown that certain species could produce culms which would constitute a good raw material for making pulp for paper. The seeds could be used as forage as they already are today. It has also been observed that the plant growth contributes to the desalination of the soil.

Another study includes species of the two genera Atriplex (oraches) and Kochia, which could possibly be developed into interesting forage plants on saline soils.

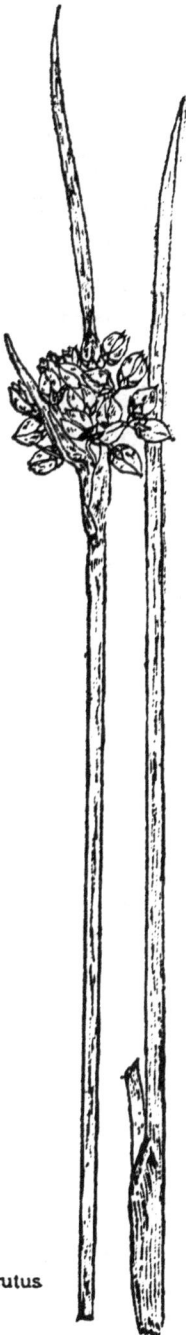

Juncus acutus

Contact: Dr Mahmoud Abdel Kawy Zahran, Professor of Plant Ecology, Department of Botany, Faculty of Science, Mansoura University, Mansoura, EGYPT.

MINI DAIRY

Christer Cronberg
Johan Bjurmar
Ingmar Kristoffersson
Alfa-Laval AB
SWEDEN

A small and flexible milk processing plant capable of producing consumer milk, cultured milk, cheese, butter etc. in a tropical or semi-tropical country. The products will be of west European food standards.

The Alfa-Laval Mini Dairy is a scaled down version of a regular milk processing plant. It can produce pasteurised, standardised and flavoured consumer milk as well as cream, cheese, butter or cultured milk such as yoghurt or labneh. The Mini Dairy has been developed for rural areas with the objective of processing as soon as possible the raw milk into more durable dairy products. The design is modular and flexible. It can handle milk not only from cows but also from sheep, goats and buffaloes. The plant can also be used for reconstitution of milk. It has been designed for tropical or semitropical climates and is delivered in a semi-assembled form. The throughput is up to 4,500 litres per 8 hours working day.

The base unit consists of tanks, pumps, pasteuriser, cooling heat pump and a cold storage space. Optional extras include e.g. cheese making tools, cream separator, butter churn and a choice of packaging equipment.

The purpose of the Mini Dairy is to preserve and upgrade a local milk supply for a not too distant small urban market. It provides an outlet for surplus milk, thus producing cash for the local dairy farmer or even for nomads, and the basis for expansion of dairy farming.

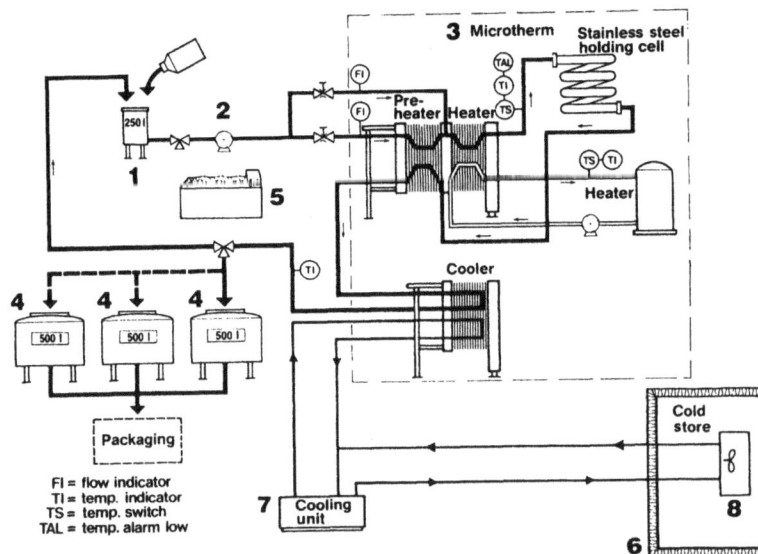

FI = flow indicator
TI = temp. indicator
TS = temp. switch
TAL = temp. alarm low

Contact: Alfa-Laval Food and Dairy International AB, Box 64, S-22100 Lund, SWEDEN. Telephone: +46-46-105000.

VILLAGE PALM OIL EXTRACTION

Outil pour les Communautés
Douala–Akwa
CAMEROUN

A set of four simple, low cost units for extraction of clarified palm oil from palm fruit bunches. In its basic form the entire plant is hand-operated and firewood is used as fuel for heating.

A complete palm oil extraction plant that consists of four units:

- Cooking drums for boiling the bunches.
- A simple hand-driven bunch stripper to separate the fruits from the stems.
- A continuous hand-driven screw press for separating oil and fruit-water from nuts and fibre.
- An oil clarifier in the form of a heated drum.

The throughput of the plant is 200 kg of bunches per hour which normally yields 34 kg of clarified palm oil. In its simplest form it requires 4 to 6 workers but it can be motorised.

The rights to use this innovation have been granted to many organisations such as GRET, ALTECH and APICA (Cameroun). It is now in use in many countries outside Cameroun for example Zaire and Guinea.

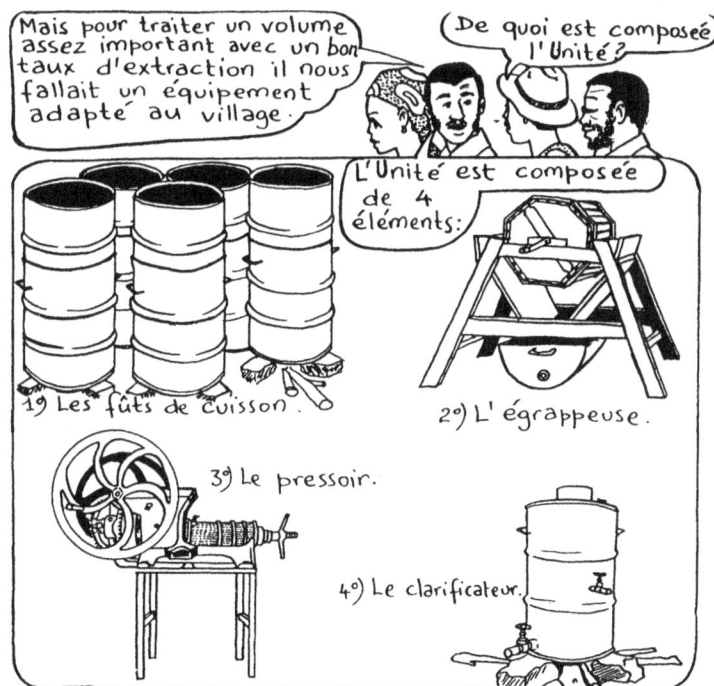

Contact: Mr Alain Laffitte, Managing Director, Outil pour les Communautés (OPC), BP 5946, Douala-Akwa, CAMEROUN. Tel: +237–421228

METHOD FOR STORAGE OF POTATOES

Dr Teófilo Jorge Aliaga Osorio
INDDA
PERU

A simple and biological method for storage of potatoes and other agricultural products at an ambient temperature. It makes use of a native South-American herb which is rich in essential oils.

This is a re-invention of a pre-Columbian Inca method for storage of potatoes. The ancient way of doing it was very simple. The potatoes were mixed with a herb called Muña and the effect was less sprouting and less damage by insects.

Recent tests have verified these effects on the potatoes and have also explained the biochemistry behind them. Muña belongs to a family called Labiatae, which is known for having many species rich in essential oils. Several of them are used in European and Oriental kitchens.

It has now been shown that potatoes can be stored at ambient temperature between harvests which saves the energy required for cold storage. Damage due to insects and sprouting was reduced and so were losses due to evaporation. Another gain was that less sugar was developed in the potatoes compared with cold storage. The R&D continues with extracts from Muña, with other species of herbs, with other tubers such as onions and also with storage of fruits such as oranges, avocados and bananas.

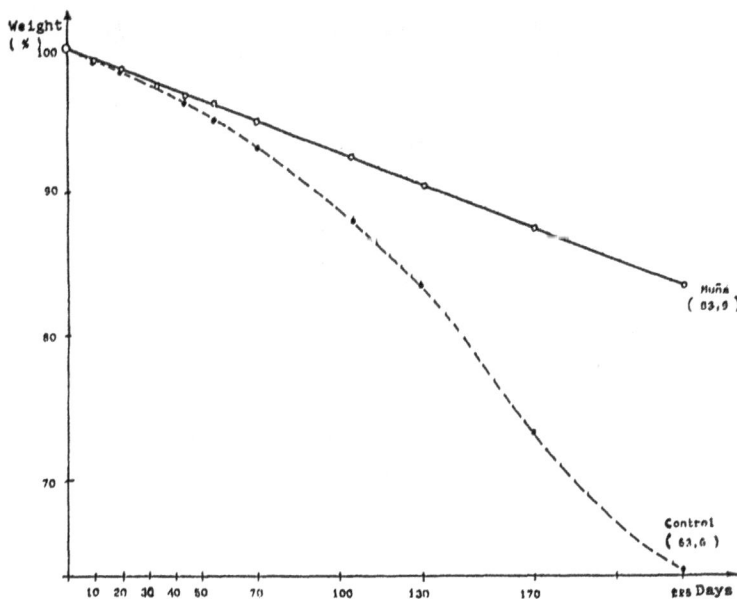

Fig. 5: Weight losses of Potatoe during storage at 20 ± 2° C.

Contact: Dr Teófilo Jorge Aliaga Osorio, Instituto Nacional de Desarrollo Agroindustrial (INDDA), Av.la Universidad 595, La Molina, Apartado 14, 0294 Lima 14, PERU. Telephone: 351630 or 353023 private.

BUILDING ELEMENTS OF ROUND-WOOD

Dr Pieter Huybers
Jaap Lanser
Mr Sier Th. van der Reyken
Delft Technical University
NETHERLANDS

Method of constructing rather Mr large buildings from round-wood of small dimensions, including a wire lacing tool for connecting and protecting the round wood ends.

The heart of this invention is a wire lacing hand-tool which makes it possible to use round-wood of small dimensions, 8 to 15 centimetres i diameter, for building purposes. Galvanised steel wire is used for the lacing which has two functions i.e. to secure connections between pieces and to avoid splitting of the ends.

The building system includes small connecting pieces of flat metal, which make it possible to use round-wood of small dimensions for building rather large house structures. In many situations this means a switch-over from imported timber to local resources. The initial investment in tools and material will be low and most of the work can be carried out locally after a short training period. The strength of the building will in most cases be better than with traditional construction methods.

The combined stretching and fixing action of the lacing tool is new. It can be used also for connecting and reinforcing bamboo pipes as well as wood stave pipes and containers.

A rather large building for testing and demonstration purposes has been erected in the Netherlands

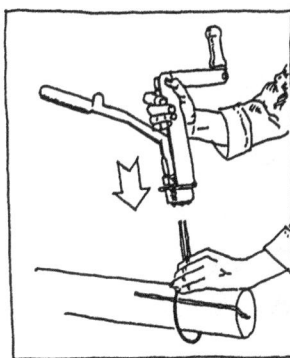

The lacing tool is being placed on the wire ends.

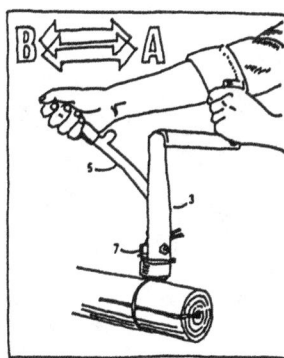

The wires are stretched by alternate movement of the handle.

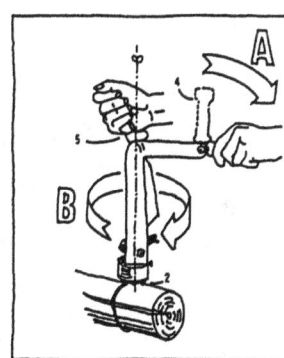

Twisting the wire ends by rotation of the tool.

Contact: Dr Pieter Huybers, Stevin Laboratory IV, P.O.Box 5048, NL-2600 GA Delft, NETHERLANDS. Telephone: +31-15-782314

WASTE PAPER MOULDING SYSTEM "MELPACK"

Mr Anthony Hopkinson
Royston, Hertshire
UK

MELPACK is a small machine for pulping waste paper and moulding it into egg trays, plant pots etc.

This innovation makes use of recycled waste paper and turns it into products to be used locally. It is cheap and reliable and occupies two operators.

The MELPACK process has five steps:

- Soaking of the waste paper
- Breaking down the paper in the pulper
- Moulding with suction on a wire mesh mould
- Blowing onto transfer moulds
- Drying on open racks

Melpack has been used for making egg trays at a rate of 60 per hour and 6-egg cartons at 180 per hour. Plant pots are made of pulp mixed with composted chicken manure. Water proof roofing tiles are among other possible products.

So far MELPACK has been built in the UK and exported to 9 developing countries. It could be manufactured locally with the possible exception of the mould. MELPACK-2 is a four times larger machine and six of this size have been sold to Africa and Latin America.

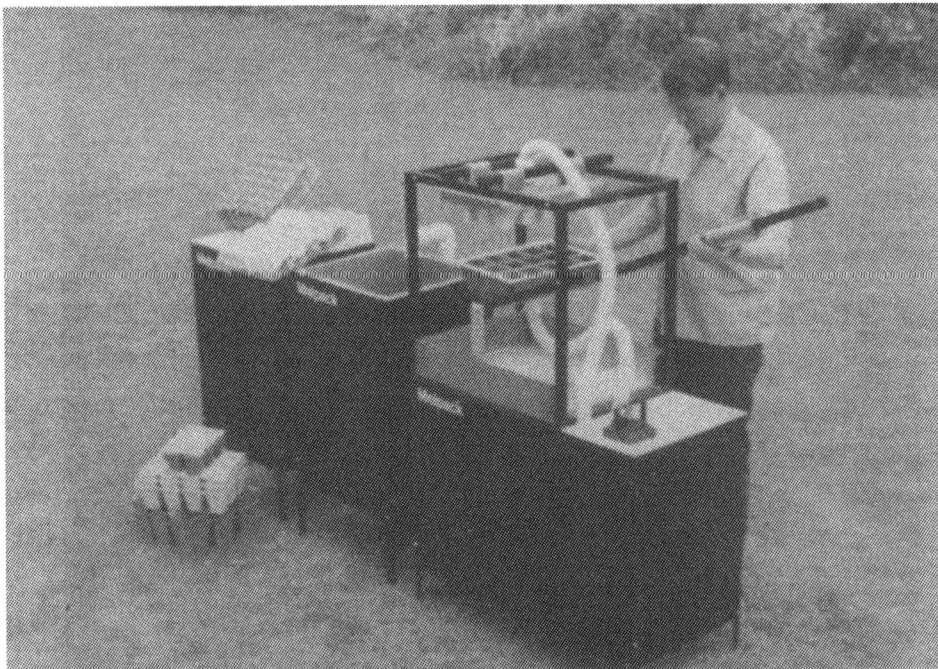

Contact: Mr Anthony Hopkinson, Third Scale Technology Ltd., Melbourn Bury, Royston, Herts SG8 6DE, UK

INCLINED CANE SUGAR DIFFUSER

Cyro Gonçalves Teixeira
José Gasparino Filho
Empresa Brasileira De Pesquisa
Agropecuária (EMBRAPA)
BRAZIL

A low cost cane sugar diffuser for extraction of sugar from cane in direct connection with a mini-distillery. It consists of an inclined trough with a screw conveyor for counter-current transport of bagasse.

A simple and low cost set of machinery for extraction of sugar from cane for production of ethanol. A cane shredder is followed by a roller mill with a single stage of rollers. The last stage is a diffuser which is the proper invention. It has two almost identical stages each consisting of an inclined trough with a screw conveyor. The latter transports the bagasse up the slope but also stirs it and thus provides the necessary lixiviation. Water is sprayed on the cane at the top of the trough where the bagasse is above the free liquid level. Steam is injected at the bottom end in order to raise the temperature to the desired 60 °C. The liquid flow is countercurrent to that of the bagasse.

A manufacturing licence has been sold in Brazil and there are now 5 units in operation. They have been easy to maintain, to clean and to inspect. The sugar yield has been above 90% which in turn means that in direct connection with a mini-distillery the yield of ethanol has been above 65 litres per ton of cane. The power consumption has been around 2.5 Hp/ton cane/hour.

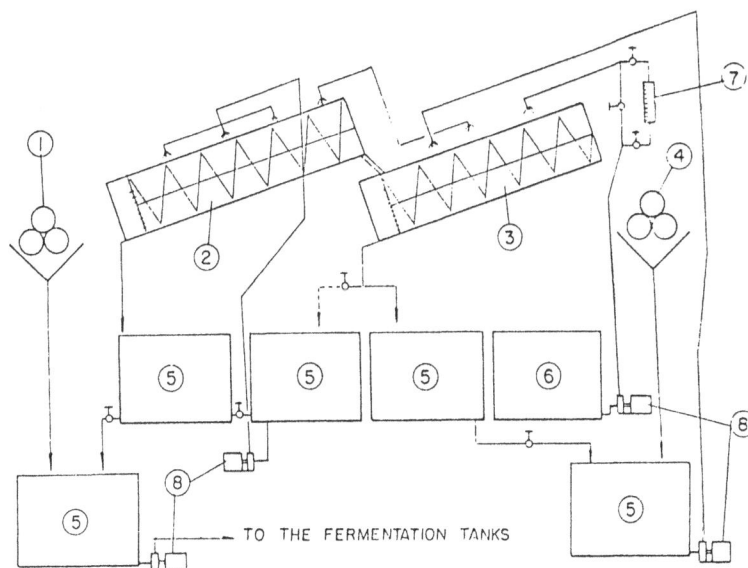

INCLINED DIFFUSER FOR MICRODISTILLERIES

— TO THE FERMENTATION TANKS

Contact: Mr Cyro Gonçalves or Mr José Gasparino, Brazilian Agricultural Research Enterprise (EMBRAPA), BRAZIL.

PALM OIL EXTRACTION MACHINE "EPOMIL"

Mr Akpan Udo Ekpo
University of Nigeria,
Nsukka
NIGERIA

A simple, cheap and easy to use machine for the processing of boiled palm fruits into palm oil and nuts. Compared with traditional methods it requires less labour and offers a good yield.

The machine consists of two main parts. The upper stationary part has a screw spindle with a press plate which is provided with a "mashing fork" in the first half of the operation cycle. The lower rotatable part has a "mortar" on a turntable and is provided with a perforated metal basket during the second half of the cycle. The table is turned at a speed of 450 or 720 rpm by a 1 Hp engine.

10 kg of boiled palm fruits are put in the mortar. The press plate with the mashing fork is lowered into position and the mortar is rotated until the nuts are separated from the pulp. The pulp is transfered to the perforated basket and is pressed by turning the screw spindle by hand. The complete cycle takes 5 - 10 minutes depending upon the skill of the operator.

The traditional methods of pounding the boiled fruits and squeezing out the oil by hand requires hard labour and gives a poor result in terms of quantity and quality. This invention, which saves labour and improves the yield, can handle all types and sizes of fruit. One prototype has been built and successfully demonstrated. Local manufacturers have shown interest but no licence has been sold so far.

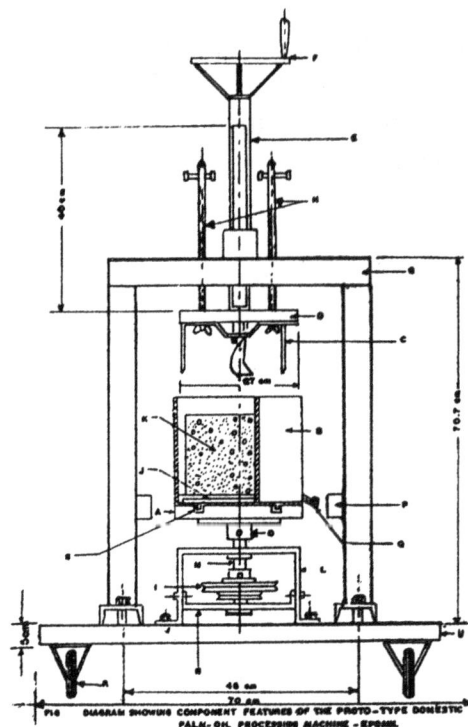

FIG. DIAGRAM SHOWING COMPONENT FEATURES OF THE PROTO-TYPE DOMESTIC PALM-OIL PROCESSING MACHINE - EPOMIL

Contact: Professor Alex O.E. Animalu, Faculty of Physical Sciences, University of Nigeria, Nsukka, NIGERIA. Tel: 6251

STABILISED COCONUT MILK

Mrs Ubolsri Cheosakul
TISTR
THAILAND

A complete process for the production of stabilised coconut milk in cans or plastic bags. The process has been put into industrial scale production in cooperation with a local food-manufacturing firm.

The inventor designed a process for production of stabilised coconut milk based on her R&D work att TISTR. The process was put into industrial production by Pua Sui Heng Ltd. (PSH) in Bangkok and has been in operation since 1980. The canned products are exported to many countries.

The process consists of the following main steps:

* Shredding of coconut kernels in a pin-roller grater. Hand feeding and a half Hp motor
* Pressing the shredded kernels in a continuous screw press provided with a 5 Hp motor. Passed three times with water added between
* Adding benzoic acid, a preservative
* Heating in a scraped surface heater to 117 °C for up to 3 minutes
* Homogenisation of lipids and precipitated protein (curd) in a high pressure homogoniser or a colloidal mill.
* Cooling and filling in cans or bags

SCHEMATIC OUTLINE OF THE PROCESS FOR PRODUCTION OF STABILIZED COCONUT MILK

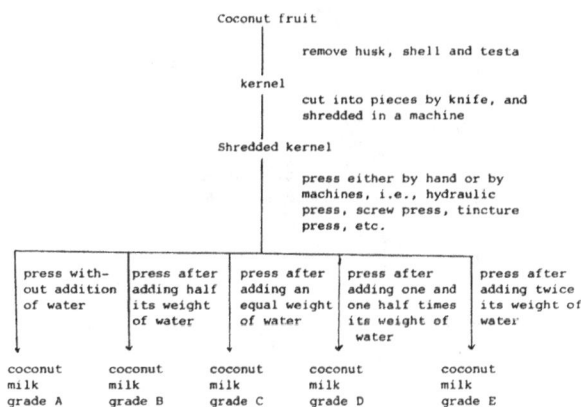

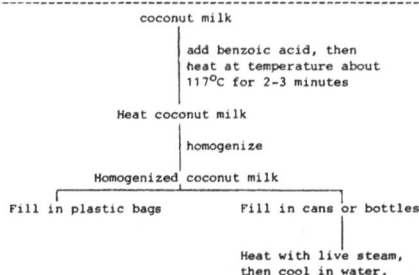

Shredded coconut or coconut milk does not keep more than a day in a tropical climate. They are basic ingredients in many Asian kitchens. Thus there is a market for stabilised coconut milk in many Asian countries and beyond.

Patent has been granted in Australia.

Contact: Prof. Smith Kampempool, Governor, Thailand Institute of Scientific and Technological Research (TISTR), 196 Phahonyothin road, Bangkok 10900, THAILAND.

WOOD STAVE SILOS AND TANKS

Mr David Henry Eadie
Ontario
CANADA

A number of methods used for joining short pieces of wood into wood-stave silos and water tanks. The invention includes the development of the required wood shaping machines.

This invention consists of a number of methods of using short pieces of waste wood to construct wood-stave silos and water tanks. It includes the development of wood-working machinery which can be manufactured by local industries.

Big containers are built from short pieces of wood which have been finger-jointed, machined to shape and impregnated with creosote or phenol against termites and other insects. The staves are kept together with mild steel rods and nuts. All the wood comes from local resources and all the work is carried out at the building site with the aid of special wood-shaping machines.

The project also includes prefabricated beams of glue-laminated wood for food storage buildings.

The post harvest losses are as high as 50% in many tropical countries. Thus safe and dry storage of food such as grain and oilseeds, is of the utmost importance.

This new technology has been tested in Burma and UNICEF has ordered 200 water tanks to be built this way.

Contact: Dr U Myint Than, Deputy Director, Cottage Industries Department, Ministry of Co-operatives, Kyanginsu Village, Mingaladon Township, Rangoon, BURMA

PLUNGER-AUGER FERTILISER INJECTOR

Dr Amir U. Khan
IRRI
PHILIPPINES

An efficient fertiliser injector for rice farmers, consisting of two skids, each provided with a container for urea, a dosage device, a furrow opener, an injection tube and two furrow closers. The nitrogen utilisation in the rice paddies is increased by 30 to 45%.

This innovation is a small and light device for injection of urea fertiliser into the soil in flooded rice paddies. Traditional methods resulted in the urea being dissolved in the water and evaporated. The present method increased the nitrogen utilisation by the rice plants by 30 to 45%

The device consists of a pair of skids and a wheel that drives the dosage of urea. Each skid has a container for the fertiliser, a furrow opener, an injection tube and two furrow closers. The complete device weighs only 7.5 kg and is pushed along with the skids between the rows of planted rice.

The fertiliser is injected at a depth of 4 cm and in such a way that there is no contact between it and the flood water. The furrow is opened immediately ahead of the auger and closed again immediately after it. Thus the urea stays in the ground where it is taken up by the rice plants. It requires about 16 manhours to inject 50 kg of nitrogen per hectare.

There are about 80 machines available for demonstrations in the Philippines. They are manufactured in 2 local workshops. The price is about 65 USD. Other specimens are being evaluated in Bangladesh, Indonesia and Thailand and there is interest in Burma, Madagascar and Vietnam.

Contact: Head of the Agricultural Engineering Department, International Rice Research Institute (IRRI), Box 933, Manila, PHILIPPINES.

METHOD OF ERECTING GRAIN-SILOS

Mr Mihály Németh
Szeged
HUNGARY

A new way of erecting large grain silos with a minimum of skilled labour and without heavy cranes.

This is a method of erecting large metal grain silos without the use of a wide flat surface for pre-assembly of ring-shaped silo segments and without the use of a heavy crane to lift the segments into place.

The first step is to build the conical bottom of the silo and to install the discharge chute and the conveyor belt for unloading of the silo. The bottom part of the silo is then filled up with screened sand or gravel with such flow properties that it can be unloaded by the grain unloading machinery after the job is done. The sand forms the working floor for the following steps in the erection of the silo.

Small prefabricated metal parts can be put directly into place without any pre-assembly outside the silo under erection. Thus there is a saving in skilled labour.

Four silos have been built in Hungary according to this method. Their capacities are 4,000, 6,000, 10,000 and 16,000 tonnes of grain respectively.

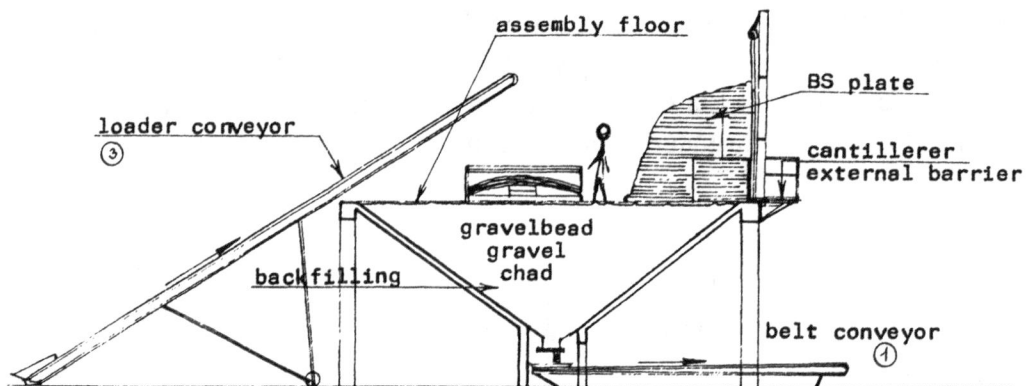

Contact: Csongrád Megyei Gabonaforgalmi és Malomipari Vállalat, PF 137, H-6701 Szeged, HUNGARY. Tel: +36-62-26622 (Csongrád County's grain-trading and flour-milling enterprises)

GARI PROCESSING PLANT

Dr Narain Das Wadhwa
Accra
GHANA

A set of machinery for the production of gari, a solid fermented food, from cassava tubers.

Gari is an important fermented food product in many parts of West Africa. It is made in several steps and this innovation consists of a set of simple machinery for village production of gari from cassava tubers. The main components are:

- Cassava peeler
- Cassava grater
- Cassava press
- Sieving machine
- Gari roaster
- Gari grading machine
- Gari grinder

Cassava tubers are peeled and washed. Clean tubers are grated and pressed. After a solid state fermentation the cake is broken up into granules, graded and toasted to crispness.

(1) Wadhwa Cassava Peeler :

(2) Wadhwa Cassava Grater :

Contact: Mr E. Abrahams, AGRICO, P.O.Box 12127, Accra North, GHANA.
Telephone: 28260

MODULAR PREFABRICATED BUILDING ELEMENTS

Mr Otto F. Joklik
Vienna
AUSTRIA

A system for making modular prefabricated building elements on the site of erection, using local wood, wall boards and cellular concrete.

This innovation is a system for the production of self supporting prefabricated building elements. It has been named BIOPOR and consists of the following operations:

A wood framework is provided with boards of gypsum, plywood or the like on one or two sides. The frame is filled with cellular (foam) concrete with a density of between 350 and 1,600 kg per cubic metre and reinforced with wood chips of defined size, shape and quantity.

The BIOPOR system is designed in such a way that the equipment for making the elements is easily transportable. Thus the elements can be produced on the building site in a "do it yourself" manner, using local raw material such as wood. A foam generator on wheels can produce up to 60 cubic metres of cellular concrete.

A license has been sold to India where 70.000 units now are being made according to a five year plan. There are negotiations with Senegal and Cameroun. The inventor is interested in joint ventures.

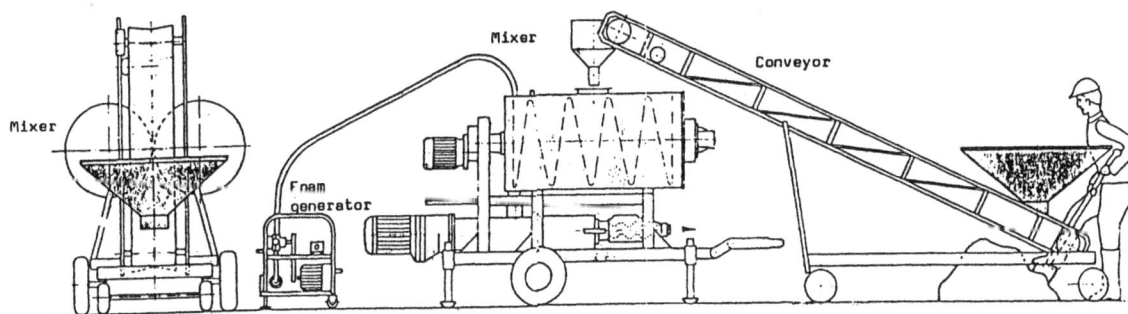

MOBILE UNIT FOR THE PRODUCTION OF CELLULAR CONCRETE (FOAM CONCRETE)

Contact: Mr Otto F. Joklik, Gersthoferstrasse 120, A-1180 Vienna, AUSTRIA. Tel: +43-222-473122

BAMBOO, WIRE NETTING AND CEMENT PLASTER WALL PANEL

Mr Gilbert A. Gibson
KITOW
VANUATU

A method of using bamboo, wire netting and cement plaster to build a low cost wall panel.

This invention is a method of making wall panels on building sites without the use of any special tools or moulds. The panels are made in four stages:

- A rectangular frame is built in wood and provided with a number of diagonal wood battons in order to give bracing strength.

- An outside layer of split bamboo, with the glossy side out, is nailed tightly side by side together.

- A wire netting is stretched inside the frame at a short distance away from the bamboo.

- The frame is filled with concrete plaster and left to set.

This method has been used successfully but rather locally on Vanuatu and neighbouring Pacific islands.

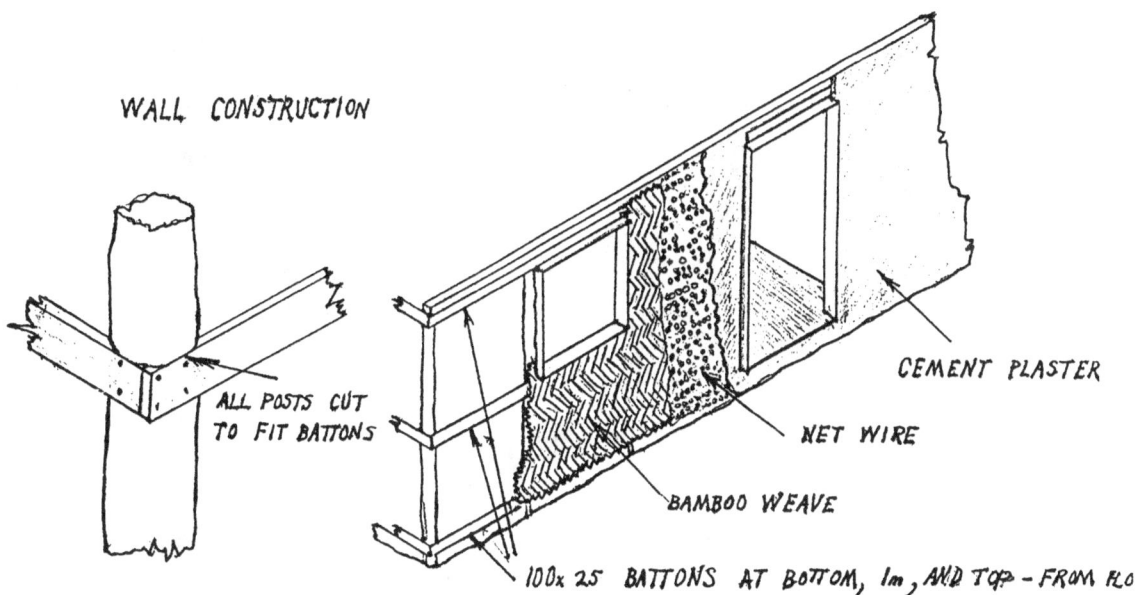

WALL CONSTRUCTION

ALL POSTS CUT TO FIT BATTONS

CEMENT PLASTER

NET WIRE

BAMBOO WEAVE

100x 25 BATTONS AT BOTTOM, 1m, AND TOP - FROM FLO

Contact: Mr Gilbert A. Gibson, Kristian Institute of Technology of Weasisi (KITOW), P.O.Box 16, Isangel, Tanna, VANUATU.

HOUSING FOR GARDENERS

Nguyen Hoang Ha
Le Kim Dung
Hoang Dinh Tuan
Nguyen Thi Hien
VIETNAM

A new type of house designed for a gardener and his family as well as his flowering plants.

This invention consists of a new type of house designed especially for gardeners and producers of flowers, where people and plants live in the same building. In the centre of the house there is a geo-thermal heating system made of bamboo and with air as heat-carrier.

The design is compact and the houses are therefore probably easy to keep warm in wintertime. The houses are intended for geo-thermal areas where the best temperature is found only two metres under the surface.

Three advantages are claimed:

• The geo-thermal system is easy to make and maintain.

• The temperature in the houses is even around the year to the advantage of man and flowers alike.

• Economy of cultivated land surface.

Contact: The Union of Architecture of Vietnam, 23 Dinh Tien Hoang, Hanoi, VIETNAM with copy to Dr. An Khang, Director General, The National Office on Inventions of the SR Vietnam, P.O.Box 432, 39 Tran Hung Dao Str., Hanoi, VIETNAM.

EXPANDING CYLINDER FOR CRACKING ROCKS

Mr Karl Gustaf Derman
K. G. Derman AB
SWEDEN

A device for cracking rocks, boulders, concrete blocks and the like consisting of a radially expandable cylinder and a high pressure hydraulic pump.

This invention consists of a device for cracking rocks, boulders, concrete blocks and the like without the use of chemicals.

The main parts of the device are a radially expandable cylinder, a hose and a high pressure pump for the hydraulic fluid. The cylinder is flexible but the endpieces are kept at a fixed distance from each other e.g. by a steel wire. The walls of the cylinder are made of an elastomer and because of the fixed endpieces the cylinder can only expand radially when pressure is applied.

A hole is drilled in the object to be cracked, the cylinder is inserted and pressure applied. The object will crack without any danger for the operators or the environment. No blasting agent or any other chemicals are used and the whole process can be carried out by unskilled workers. The system has been named DERMANITE and should be suitable for a developing area.

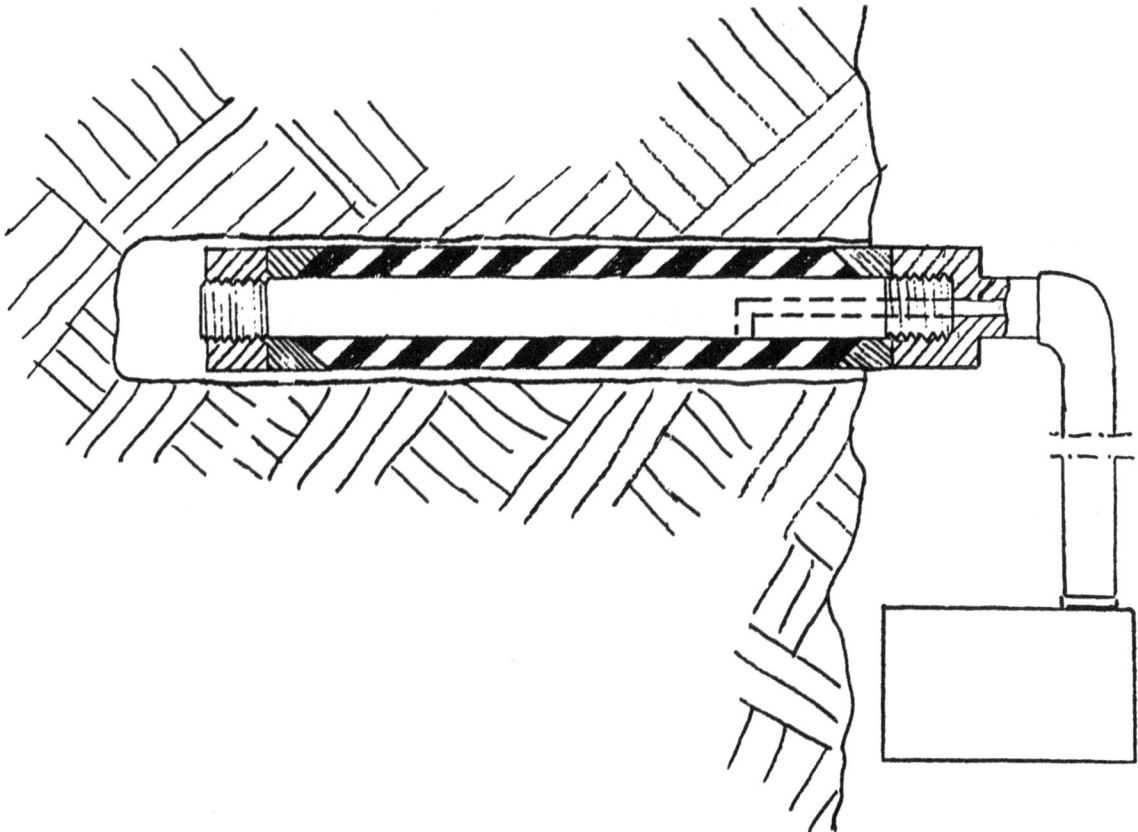

Contact: Mr Karl Gustaf Derman, Idéutveckling K. G. Derman AB, Sörgårdsvägen 7, S-433000 Partille, SWEDEN.
Telephone: +46-31-264768

MOULD FOR MODULAR CONCRETE BUILDING BLOCKS

Mr Ellis Owen Jones
Thames
NEW ZEALAND

A hand mould for making modular building blocks for cyclone proof buildings on Pacific islands.

The background to this invention is the fact that the islands in the Pacific often are struck by cyclones that leave no buildings standing. Thus there is a need for constructing at least the kitchens and the water supplies in cyclone proof material.

Modular building blocks are made of mainly local material such as sand and coral rock. They consist of steel reinforced hollow concrete. They are made in the village by the villagers with the aid of a light modular hand mould that can be rapidly converted to make all essential block types needed for even quite complex and large buildings.

The system has now been tested in the field for several years and it has been proved that:

- The buildings are permanent and the building technique is spreading from village to village.

- The buildings are well suited to the environment

- 80% of the material, i.e. sand and coral rock aggregate, is in plentiful local supply without cost

- The making of the concrete blocks and building with them has become a village industry totally integrated into the village way of life.

This building system is now used in a number of Pacific Island groups at a rate of about two buildings per village and year. Both male and female labour is employed and the investment in tools and imported material is very small.

Contact: Mr Ellis Owen Jones, Oakley Crescent, Thames, NEW ZEALAND 87991. Telephone: 89400

THE JUTTON PROJECT

Dr Mohammed Siddiqullah
BCSIR
BANGLADESH

Jutton is the name given to a jute fibre which has been modified in such a way that it has been given practically the same spinning properties as cotton.

The essence of this innovation is a physico-chemical modification of jute that gives the fibre spinning properties similar to those of cotton. The modified jute fibre has been named "Jutton".

Standard commercial cotton textile machines have been used for spinning jutton alone or in any mixture with cotton. Cloth has been made from e.g. 35% jutton and 65% cotton and made up into clothes and bedsheets. Durability tests on clothes have given results equal to and up to 50 % better than cotton. Jutton has been tested also on the markets in USA and Japan.

A pilot plant was started up in 1984 and has produced jutton for various tests. Cloth has been produced by local weavers using simple methods. This is a rather conservative market and development is slow.

Jutton has also been tested successfully in mixtures with polyester, viscose and acrylic artificial fibres.

Spinning of jutton yarns under the supervision of
Dr. Siddiqullah, inventor scientist (left).

Contact: Dr Mohammed Siddiqullah, Jutton Project Director, Bangladesh Council of Scientific and Industrial Research (BCSIR), Dhanmondi, Dhaka-5, BANGLADESH

WINDBREAK SCREEN WALL

Mr M'backé Niang
Dakar
SENEGAL

A building block to be used for the construction of screen walls against sandstorms, rain and sunshine.

This invention consists of a building block of special design. It is a type of large brick that lets through air. When made into vertical walls it provides ventilation but at the same time acts as:

- Protection against dust- and sand-storms. It provides a windbreak but does not accumulate dust in any quantities and is easy to clean.

- Protection against sunshine. It provides coolness and shade and at the same time privacy.

- Protection against rainstorms.

The building elements can be made of ceramic materials or of metals.

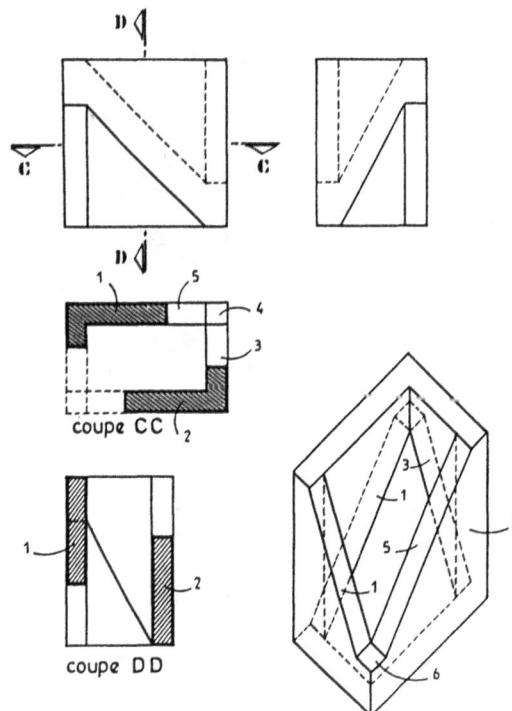

coupe CC

coupe DD

Contact: Mr M'backé Niang, Architecte DESA, B.P. 6072, Dakar, SENEGAL.

CULTIVATION OF OYSTER MUSHROOMS

Dr Arun Vinayak Sathe
MACS Research Institute
INDIA

A modified and highly efficient method of cultivating oyster mushrooms in Indian villages, with straw as the basic raw material.

In certain parts of India there is a long tradition of collecting wild mushrooms (Agaricales) in the forests surrounding the villages. This activity is not very efficient and is furthermore hard on the forests.

Oyster mushrooms of the genus Pleurotus are easier and cheaper to cultivate than the usual white button mushroom. Development work at MACS has shown that the costs for oyster mushroom production can be reduced to about half of those encountered when using conventional methods.

A typical batch will yield 900 grammes of fresh mushrooms in the first flush after 25 days. A second flush will give another 400 g after 10 days and a third flush will give 200 g after 15 days. The total yield is thus 1.500 g in 50 days. The yield in terms of fresh mushrooms/dry straw is about 2/3

Data from extensive field trials are being evaluated. The next steps of the project are to describe the process and to broadcast the information in the villages.

There are no patents on the process.

Contact: The Director, Maharashtra Association for the Cultivation of Science (MACS), Research Institute, Law College road, Pune 411 004, INDIA. Telephone: 56357

DEVELOPMENT OF LEUCAENA
PRIZE WINNERS

Mr Michael Benge
Dr. James L. Brewbaker
Dr. E. Mark Hutton
USA and AUSTRALIA

Development of the tree Leucaena leucocephala for reforestation, soil stabilisation, nitrogen fixation, windbreaks and firebreaks as well as for the production of feed for ruminants, fire wood and timber.

This innovation consists of development of a tree species for a variety of uses in tropical and subtropical countries. The scientific name is Leucaena leucocephala and it belongs to the family Leguminosae which is known to comprise many species able to fix nitrogen from the air. Leucaena can do this at a rate in the order of 500 kg/hectare/year. Thus normally no fertilisers are required.

Some varieties of Leucaena are extremely fast growing and 6 metres in the first year has been recorded. After cutting the stems, the stumps produce new shoots and no replanting is required. Thus the costs for cultivating Leucaena are low both in terms of money and labour and it is very suitable for reforestation projects.

Leucaena stabilises the soil and prevents erosion from wind or water. This tree has deep tap roots and stays green also in the dry season. Thus it also provides shelter against sunshine, forest fires and high winds. It can also serve as a living fence. Leucaena produces a foliage of high nutritional value as feed for ruminants. It contains 27-34% proteins and is both palatable and digestible.

Large scale plantation schemes have been undertaken e.g. in the Philippines, Taiwan, India, Haiti, Kenya, Mexico, Thailand and many other countries. New varieties and technologies are continuously being developed.

Transplanting of bagged leucaena seedlings.

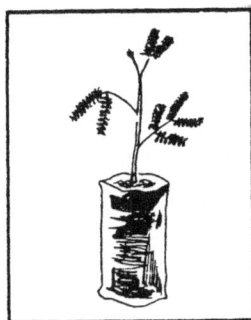

Seedlings may be transplanted when 20-50cm tall

Remove seedling from the bag being careful to keep soil around roots intact

Place the seedling upright in the hole so the root collar is level with soil surface

Contact: Mr. Michael Benge, Agency for International Development, Washington, DC 20523, USA

BULLET TREE-PLANTING

Prof. John Walters
Shropshire
England UK

A mechanised tree-planting system based on "bullets", which consist of rigid plastic plant pots made to narrow specifications. Each bullet is made of four identical pieces, which after planting are forced apart by the growing tree roots.

This tree-planting system was developed in British Columbia to be used in reforestation and nursery operations. The seedlings are grown in rigid plastic pots, which are called bullets because of their shape and because they are injection planted with tools called guns.

Figure 3. — Single-shot gun showing insertion of first bullet on strip of twelve 4½-inch bullets. Bullet is cut off the strip by depressing the handle. The bullet then drops to the bottom of the gun where it is planted by depressing the foot-bar.

Each bullet is made of four identical pieces which fit together and make up a pot with a square section and a conical bottom. The pieces are injection moulded to very narrow specifications and strapped together to form blocks of 54 pots, which are filled with soil and seeded. The seedlings in their bullets are injection planted with the aid of a manual planting gun or a more sophisticated machinery on wheels called the ReForester. After planting, the growing roots of the young tree will force apart the four pieces of the bullet. Furthermore the plastic used contains wood-waste and is biodegradable.

The advantage of the system lies in the possibility it offers to plant a large number of trees at a fast rate without any damage to the roots of the seedlings. The system is widely used in Canada.

Contact: Prof. John Walters, Beech Cottage, Deerfold, Lingen, Bucknell, Shropshire SY7 0EE, UK

SAWBLADE FOR ABRASIVE HARDWOODS

Dr Alexandre Krilov
Forestry Commission
N.S.W
AUSTRALIA

A new bandsaw tooth profile which has increased cutting speeds, extended blade life between saw changes and reduced saw kerf when sawing refractory hardwoods.

This invention has been developed within the Wood Technology and Forest Research Division of the Forestry Commission of New South Wales.

The wood species of tropical and subtropical countries often has a high content of silica which make them very abrasive. Thus the lifelength of sawblades is much reduced. This is a factor of importance in many developing countries.

The key to success has been a well designed and properly hardened saw tooth and is the result of practical as well as theoretical work including tests on e.g. silicious wood from the species Tristania conferta

The new bandsaw tooth profile and hardening has increased cutting speeds by at least 45%, extended blade life between saw changes by at least 290% and reduced saw kerf by 10% in practical trials with silicious and thus abrasive hardwoods.

Patents have been applied for in 9 countries and in Europe. The invention has not yet been taken up commercially.

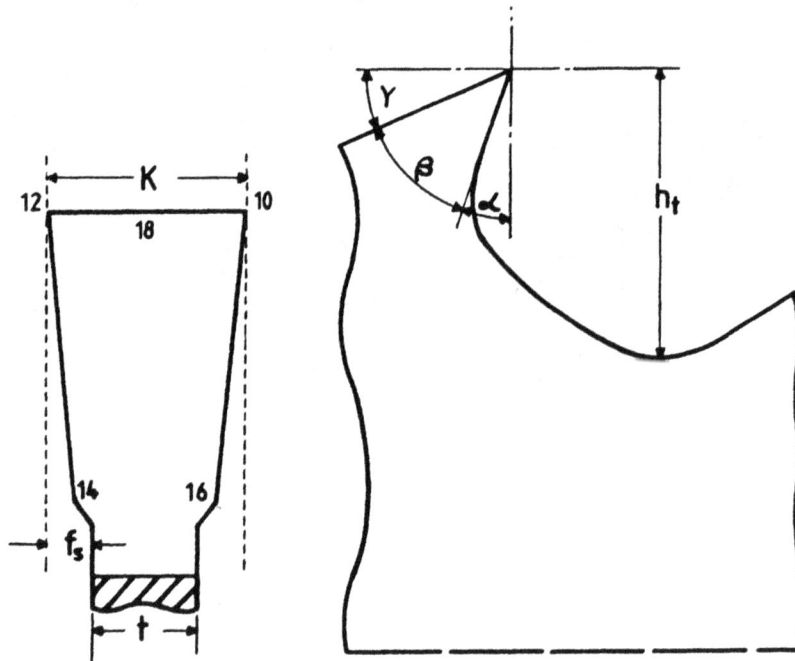

Contact: Mr L. S. Mors, Secretary, Forestry Commission of New South Wales, G.P.O. Box 2667, Sydney, N.S.W. 2001, AUSTRALIA.
Telelephone: +61-2-2341631

MYCORRHIZA TECHNOLOGY

Dr Reynaldo E. de la Cruz
UPLB College of Forestry
Laguna
PHILIPPINES

A system of collecting spores from mycorrhiza-forming fungi, of preserving the spores and of inoculating the tree seedlings, resulting in improved growth rates and savings in fertilisers.

Mycorrhiza is the name given to fungi which live in symbiosis with the roots of trees. They contribute to the health of the host by fixing nitrogen from the air and by facilitating the uptake of mineral nutrients from the soil.

The present invention is a system for inoculating tree seedlings in a nursery with spores of Mycorrhiza. Mature fruiting bodies of the appropriate fungal species are collected in season in established and healthy tree plantations. The spores are cleaned by sieving and freeze-dried or oven-dried on charcoal. The dried spores are put into tablet form and inserted into the soil of the pots in the nursery when the new seedlings are first transplanted. Other ways of applying the spores have been tested. Mycorrhiza inoculation is very important in reforestation projects on "virgin soil" and most important when introducing new species of trees. The present invention was tested in the Philippines on introduced species of pine and gumtree. In an ANZAP nursery Pinus caribaea grew 108% taller and 58% thicker after inoculation with Mycorrhiza spore tablets. Corresponding figures for Eucalyptus camaldulensis was 20% and 35%. The saving in fertiliser was 70%. This invention can be operated with very simple means. The tabletting machine can be manual and the costs per tablet kept very low.

Table 2. Height and diameter responses of Eucalyptus and pine species to inoculation with mycorrhizal tablets in various cooperator nurseries.

| | PINES | | | | EUCALYPTUS | | | |
| | HEIGHT | | DIAMETER | | HEIGHT | | DIAMETER | |
	cm	% increase	cm	% increase	cm	% increase	cm	% increase
		ANZAP P. caribaea				ANZAP E. camaldulensis		
INOCULATED	36.7	108%	0.30	58%	66.4	20%	0.35	35%
UNINOCULATED	17.6		0.19		55.4		0.26	
		NIA P. kesiya				NIA E. camaldulensis		
INOCULATED	25.5	15%	0.26	13%	38.7	22%	0.31	11%
UNINOCULATED	22.2		0.23		31.7		0.28	
		CRC P. kesiya						
INOCULATED	10.6	28%	0.15	17%				
UNINOCULATED	8.3		0.13					

Contact: Dr R.E. de la Cruz, Department tef Forest Biological Sciences (FBS), UPLB College of Forestry, College, Laguna, PHILIPPINES.

PLASTIC SLEEVE TREE SHELTERS

Mr Graham Tuley
Kincardineshire
SCOTLAND

Plastic sleeves of various designs and materials of construction to protect young trees from browsing animals and to create a microclimate favourable for fast growth.

This innovation consists of a plastic sleeve which is put around planted or naturally regenerated tree seedlings. It is made of clear translucent plastic with an expected life of at least 5 years. The primary function is to protect the young tree from mammals such as rabbits, deer, cattle and horses. Many different designs are offered but they all have in common that they need a form of support. They vary in height from 0.6 metres (rabbits) to 2.0 m. (cattle and horses).

The sleeves have improved the survival of the trees. Another advantage is that chemical control of competing vegetation has been made easier outside the shelter. The sleeves have been found to result in increased growth for all broadleaved species and most conifers. This is probably due to a favourable microclimate inside the sleeves.

About 100,000 sleeves per annum are now used in UK.

There is no patent on the initial invention.

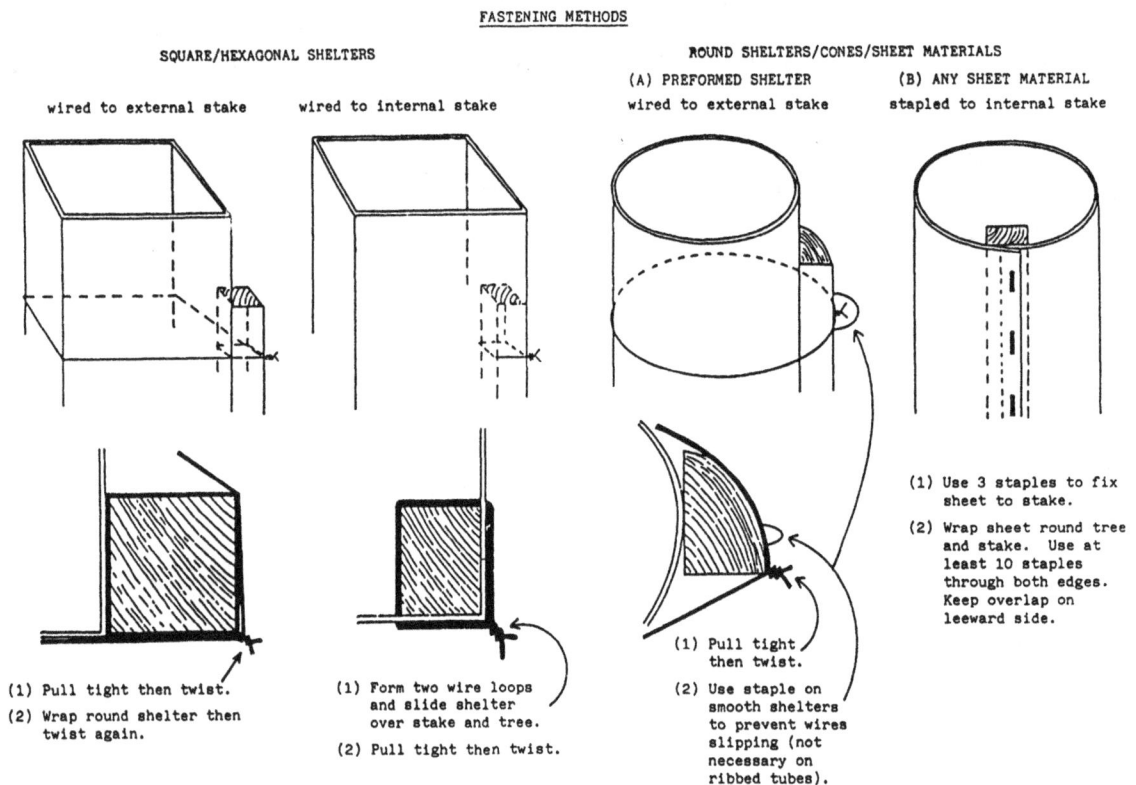

FASTENING METHODS

SQUARE/HEXAGONAL SHELTERS

wired to external stake   wired to internal stake

ROUND SHELTERS/CONES/SHEET MATERIALS

(A) PREFORMED SHELTER
wired to external stake

(B) ANY SHEET MATERIAL
stapled to internal stake

(1) Pull tight then twist.

(2) Wrap round shelter then twist again.

(1) Form two wire loops and slide shelter over stake and tree.

(2) Pull tight then twist.

(1) Pull tight then twist.

(2) Use staple on smooth shelters to prevent wires slipping (not necessary on ribbed tubes).

(1) Use 3 staples to fix sheet to stake.

(2) Wrap sheet round tree and stake. Use at least 10 staples through both edges. Keep overlap on leeward side.

Contact: Mr Graham Tuley, Tuley Tubes Ltd., 5 Murray Place, Stonehaven, Kincardineshire AB3 2GG, Scotland, UK. Telephone +44-569-65181

RAPID GERMINATION OF RATTAN SEEDS

Ms Antonia L. Agmata
Forest Res. Inst.
Laguna
PHILIPPINES

Rattan seeds germinate very slowly, which is an obstacle when cultivating this important tropical plant. By removing the hilar cover of the seeds the time for germination was brought down from over 90 days to only 2.

Rattan is the name for a number of climbing palms which grow in the tropics. They are widely cultivated but the slow germination of the seeds limits the extension of the plantations

Rattan provides material for furniture, baskets, containers and other handicraft. It is locally an important crop and a source of export revenue.

This innovation consists of removing the "hilar cover" of the fresh seeds with a knife. The result is striking. In one species called Palasan (Calamus merillii) the germination time at ambient temperature was reduced from between 90 and 210 days to only 2 days. The very simple operation can be carried out by unskilled labour at a rate of 10 seeds per minute.

Hilium = a scar on a seed, marking the place where it was attached to the seed stalk (Webster).

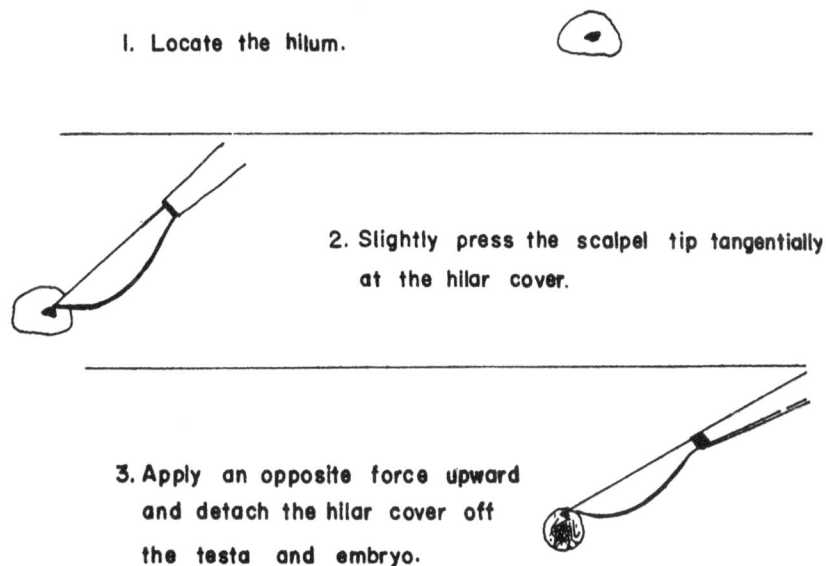

1. Locate the hilum.

2. Slightly press the scalpel tip tangentially at the hilar cover.

3. Apply an opposite force upward and detach the hilar cover off the testa and embryo.

Stages of hilar cover removal of Calamus merrillii Becc.

Contact: Forest Research Institute, College, Laguna, PHILIPPINES.

ELECTRONIC LOAD CONTROLLER FOR MICRO HYDRO ELECTRIC PLANTS

PRIZE WINNERS

Mr Rupert Armstrong Evans
Mr Gerald Pope
ITDG
UK

A simple and reliable electronic method of controlling the frequency of the alternating current from a generator which is driven by a small water turbine.

In many developing countries there is no extensive electric grid and thus a great need for small decentralised hydro power plants. The electrical output from these has to have a precisely controlled voltage and frequency in order to fit standard motors for alternating current.

This innovation consists of a sensor that measures the frequency of the electrical output from the generator. An electronic control device deviates any surplus electrical energy from the regular load to a parallel resistive heating ballast. The result is a system which is cheap to install and maintain, simple and reliable and fast enough to protect standard motors.

The load controller has been installed in more than 20 countries and some have been in operation for more than 11 years without failures. Licences have been given to Thailand and China. The manufacturing capacity in UK is now 30 devices per year in a range up to 100 kW.

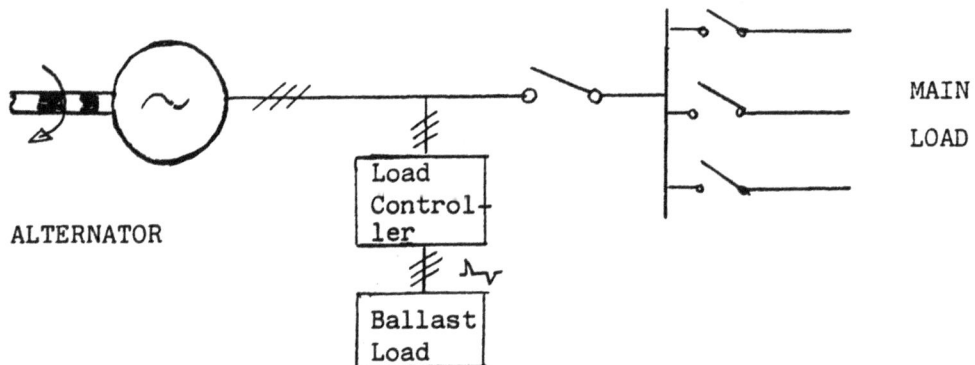

Contact: Mr Gerald Pope, G.P. Electronics, Pottery Road, Bovey Tracey, Devon, UK. Telephone: +44-626-832670

THE INTERMEDIATE TECHNOLOGY WINDPUMP

PRIZE WINNER

Mr Peter L. Fraenkel
I. T. Power Ltd.
Berkshire
UK

A windpump for water supplies and irrigation systems designed for low manufacturing costs and low maintenance as well as high efficiency and reliability. It is made by local small industries e.g. in Kenya and Pakistan.

Windpumps were an essential part of many agricultural frontiers in e.g. USA and Australia. This innovation is a modernisation of the design in order to make them more suitable for local manufacturing in developing countries.

The project includes a manufacturing manual as well as an installation manual. The rather few components are prefabricated locally except the roller bearings. The parts are assembled at the well on the ground. The tower is hinged and is raised after completion. It can be lowered again for bore hole maintenance. The cantilevered blades are slightly twisted for optimal aerodynamics. When the wind speed is too high the tail and the rotor folds together and the rotation stops. At lower winds it starts again automatically. Maintenance is reduced to greasing of the bearings once a year. The predicted life length is 20 years or more.

4 sizes have been built so far. The rotor diameters are 3.7, 4.9, 6.0 and 7.5 metres. The weight of the complete larger mills is 1.200 and 1.400 kilogrammes respectively. There are no patents to protect this innovation. A complete package is offered on license from I.T.Power. The pump is now manufactured in Kenya and Pakistan and prototypes are on demonstration in many other countries. About 110 pumps are in operation, some of them for more than 7 years.

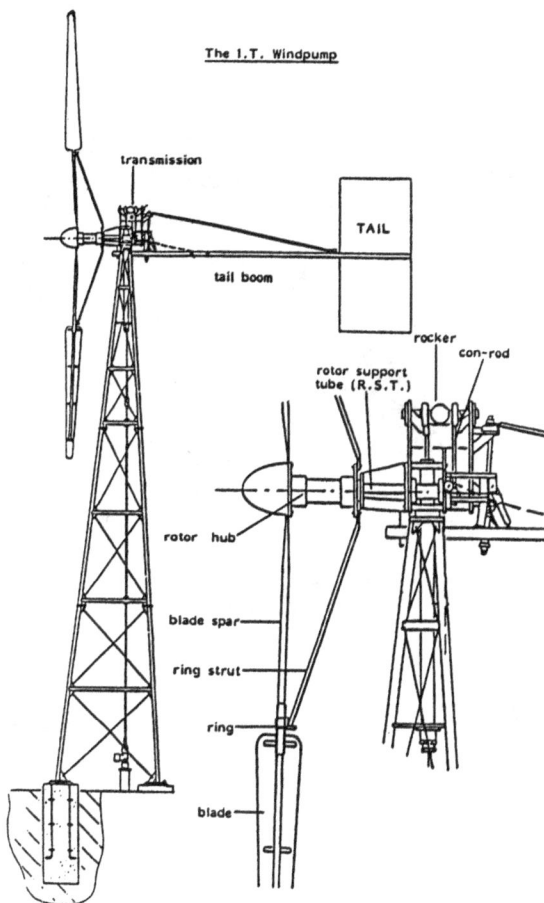

The I.T. Windpump

transmission

TAIL

tail boom

rocker
con-rod

rotor support tube (R.S.T.)

rotor hub

blade spar

ring strut

ring

blade

Contact: Mr Peter.L. Fraenkel, Intermediate Technology Power Ltd., Mortimer Hill, Mortimer, Reading, Berkshire RG7 3PG, UK.
Telephone +44-734-333231

WOOD FIRED COOKING STOVES

HONOURARY AWARD WINNERS

Mrs Margerite Kabore
Mr Rigobert Yameogo
Ouagadougou
BURKINA FASO

A number of new designs for low cost cooking stoves made in ceramics or sheet metal and fired with wood or charcoal.

This innovation is a complete national programme for providing Burkina Faso with more efficient cooking stoves.

The reason for the programme is the shortage of wood, which results in deforestation and a rapidly advancing desert.

The programme has produced a number of low cost high efficiency stoves of various designs. They are all fired with wood or charcoal and they all take ceramic cooking pots. The saving in energy is said to be 40 to 50 %

The stoves are made of ceramics or sheet metal. In the first case the raw materials are local and the stoves can be made by a local potter. In the second case the raw material may be recycled scrap sheet metal which can be put together without welding by a local blacksmith.

The development of the new technology is now followed up by an information, demonstration and distribution programme in all the provinces of the country. Thousands of stoves of the new type are already in use in the villages.

PLUS DE CUISINE SANS FOYERS AMELIORES
POUR UN BURKINA VERT

LES FEMMES BURKINABE

Utilisent

LES FOYERS AMELIORES

Contact: Mrs Margerite Kabore, Chef du Service Foyers Améliorés, Ministere de l'Environnement et du Tourisme, BP 07044, Ouagadougou, BURKINA FASO.

BAGASSE DEWATERING PROCESS
BAGATEX-20

HONOURARY AWARD WINNER

Mr Alexandre Aidar J.
Usina Santa Lydia
BRAZIL

A process for upgrading of bagasse consisting of inoculation and baling of bagasse followed by solid state fermentation and open shed air drying. The bagatex product has a better fuel value and can be stored between campaigns.

Cane sugar bagasse is an important fuel for the sugar mill itself but is bulky and has a low energy density. Furthermore it can not be stored for more than 2 months without serious deterioration.

This invention increases the energy density of bagasse about 5.6 times i.e. to a net heat content of 3,250 kcal/kg. Bales of 170 kg with a density of 375 kg per cubic metre can be stored for more than 24 months without problems. The treated bagasse is a good raw material for production of pulp, paper and board.

We have no satisfactory technical description of the process but it seems to consist of the following steps: The natural bagasse is inoculated with microorganisms and pressed into bales. A high temperature solid state fermentation converts sugars into organic acids. This in turn makes it possible to dry the bales in open sheds to a moisture content of about 20%. The whole process takes 20 days. The dewatered bagasse can be stored in the open under tarpaulins.

Three Bagatex-20 plants are in operation in Brazil. Another plant has been exported.

Characteristics of the Product Obtained in the Existing Plant

Bagass-20 currently produced at Usina Santa Lydia pilot plant possesses the following characteristics:

- moisture content:	20%
- LHC (Lower Heat Content):	3,244 kcal/kg
- bale dimensions:	1.10m x 0.90m x 0.55m
- average bale weight:	170 kg
- density:	312.2 kg/m³ (20% moisture)
- packing:	8-bale pallets

Contact: Mr Alexandre Aidar, Usina Santa Lydia S.A., Bodovia Mario Donega km 2, Caixa Postal 58, Riberão Preto, CEP 14000 BRAZIL. Telephone: +55-16-634-4030

MULTIPURPOSE WATERPOWER UNIT

Mr Akkal Man Nakarmi
Mr Andreas Bachmann
Kathmandu Metal Industry
NEPAL

A waterpower unit consisting of a horisontal water wheel with a belt pulley that can drive a variety of farm machinery such as flour mills, oil expellers etc. The metal parts are delivered as the "MPPU Construction Kit" and the farmer builds the rest.

The MPPU (= Multi-purpose power-unit) Construction Kit contains all the metal parts required for building a horisontal water wheel provided with a belt pulley for driving a variety of farm machinery. All parts are simple in design and can be manufactured in a local workshop. The farmer/buyer uses local sources of wood and stone and adds his own labour to build the frame, the dam, the pipes etc. that are required for the complete power unit.

The traditional design with a millstone on the same shaft on top of the water wheel results in a very heavy load on the bottom bearing.

The MPPU Construction Kit offers many advantages. It is cheap. It is efficient and in one case the output of flour milling was increased four times. The buyer can contribute with his own labour and can use local materials of construction such as wood and stone. All parts are portable which is very important in those hilly parts of the world where this kind of water power is available. The unit is flexible and can be used for driving many different kinds of farm machinery. With several sizes of driving pulleys available the driving belt speed can be adapted to the demands of the specific machine. More than 110 units are in operation in Nepal, Bhutan and India.

MPPU has been developed with assistance from UNICEF.

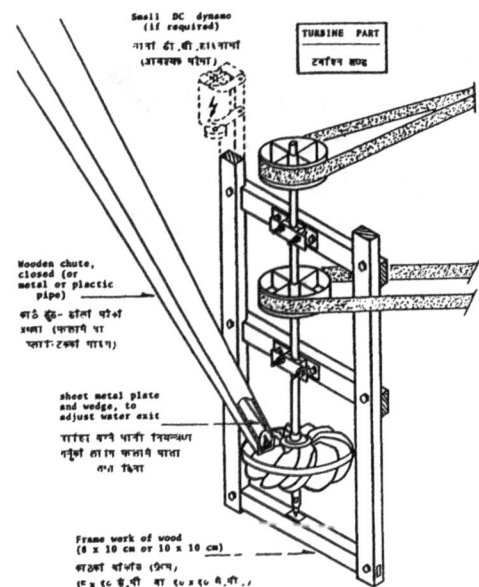

Contact: Mr Akkal Man Nakarmi, Kathmandu Metal Industry, 12/514 Quadron, Nagal Chhetrapati, Kathmandu, NEPAL. Tel:0214069

PORTABLE SOLAR TIMBER KILN

Mr R.A. Plumptre
Commonwealth Forestry
Institute
Oxford University
UK

A portable solar kiln to be used in combination with air drying of timber. It consists of an aluminium frame covered with a sheet of plastic, an inner heat absorbing surface and a pair of electrical fans for air circulation.

In tropical or subtropical countries air drying of timber often is difficult and results in spoilage. Forced drying is too expensive and may result in cracking and deformation. This innovation caters for e.g. furniture makers in developing countries who need a seasoning that results in wood stability and conservation.

The kiln consists of a frame of aluminium covered with a sheet of "Melinex", which is the ICI trade name for a clear plastic that allows heat radiation to pass through and is predicted to last in tropical conditions for 4 to 5 years. An inner roof of corrugated metal painted matt black absorbs the heat. Two electrical fans, 1/4 Hp each, circulate the air in the kiln, which has a capacity of 7 cubic metres.

The best economy is achieved by starting with conventional air drying of the freshly cut and stacked timber. Without restacking it can be finished off in this solar kiln.

This innovation offers low capital and operating costs and requires no particular skills from the operators. The waste is considerably reduced compared with conventional methods. In Sri Lanka good quality was obtained after 6 to 10 days of air drying followed by the same time in the kiln.

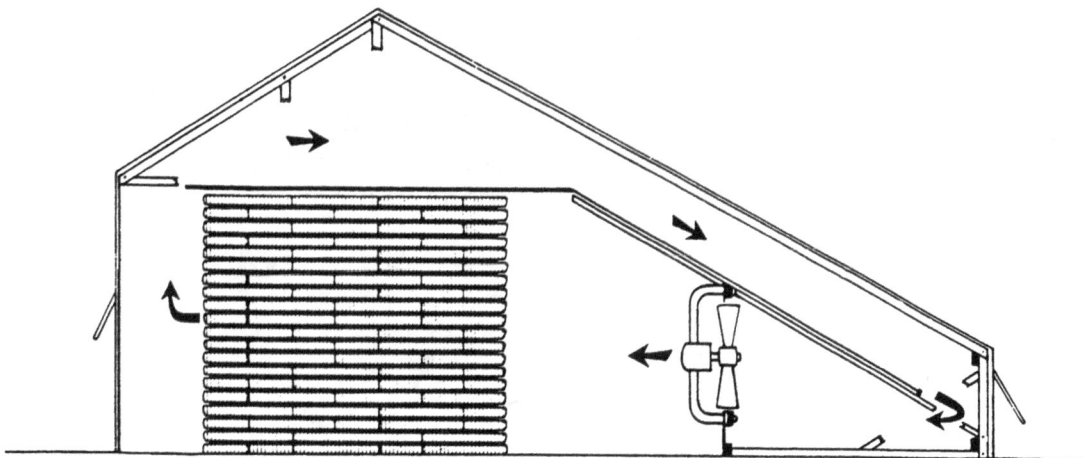

SCHEMATIC DIAGRAM OF SOLAR TIMBER KILN

Contact: Cambridge Glasshouse Co. Ltd., Comberton, Cambridge CB3 7BY, UK.

WIND-DRIVEN ICE PRODUCTION

Rui Spencer Lopes dos Santos
ENACOL
CAPE VERDE

A system consisting of a hydraulic pump which drives a hydraulic motor and the heat pump in an ice-machine. The system permits the necessary control of the speed of the ice-machine's compressor.

The purpose of this invention is to produce ice for the fishing and food industries on coasts and islands with a steady supply of wind energy.

All the main components of the system are cheap and easily available and the whole ice-plant can be put together and maintained locally. It consists of a wind wheel with a diameter of 3 metres. This drives a piston pump which delivers oil with a pressure of 70 kg/sqcm (7,000 kPa). The oil drives a hydraulic motor which in turn drives the compressor of an ice machine (400 Watt, 700 rpm). The system is designed for an average wind speed of 6 m/s. The purpose of the hydraulic power transfer is to make it possible to by-pass a certain amount of hydraulic oil and thus regulate the speed of the compressor to its optimum.

The inventor points out the possibility of building a "windfarm" which delivers pressurised oil to a big centrally situated ice-factory.

This system offers safety and a steady performance. It also offers flexibility. Surplus power, or alternatively all the power from the wind wheel can be used for driving other machinery e.g. in a fish processing plant. There are also alternative uses for the heat pump such as desalination of water through evaporation.

The feasibility of the system has been demonstrated with a prototype, which has produced ice-bars of 6 kilogrammes a piece and with a temperature of minus 11 degrees C. The development was carried out in the village of São Pedro on the island of São Vicente, where the direction of the wind is almost constant and the speed of the wind is about 8 metres per second or more.

Contact: Mr Rui Spencer Lopes dos Santos, Empresa National de Combustíveis ENACOL, Apartado 1, São Vicente, CAPE VERDE.
Telephone: 2615, 2616, 2627

WATER-DRIVEN HEAT-GENERATOR

Mr Reinhold Metzler
Furtwangen
FRG

A hot air generator which consists of an "inefficient" fan in which turbulence and internal friction in the air convert mechanical energy into heat.

This innovation is a device which converts mechanical energy into heat. It was designed primarily for mountainous countries like Nepal where there is plenty of water power but a shortage of fire wood.

The heat generator resembles a fan with baffles on the inside of the stationary housing. The baffles create turbulence and internal friction in the air. A throttling valve at the outlet and an internal recirculation valve control the amount of air delivered and its temperature. The maximum temperature achieved so far is 250 degrees centigrade.

A special boiler or kettle has been designed and tested. It is provided with fins on its cylindrical parts in order to improve the heat transfer from the hot air to the liquid in the boiler

The heat generator is intended for drying, boiling and distilling operations. The first demonstration plant in Nepal is a mill for handmade native paper in which the hot air is used for boiling the pulp and for drying the finished paper sheets.

The heat generator can be made in a local workshop. It substitutes fire wood which is a scarce commodity in Nepal. Similar conditions are found in Peru and Bolivia.

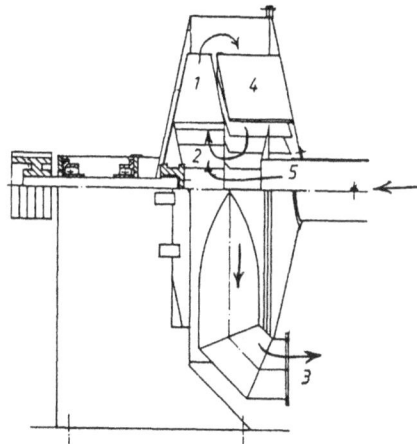

The heat generator

Contact: Mr H. Milcke, Fachstelle für Kontextgerechte Technik, Gänsheidestrasse 64, D-7000 Stuttgart 1, FRG.

GREEN BIOMASS FUEL DENSIFICATION

Watna Stienswat
Vitawas Buachandra
Kasetsart University
THAILAND

A system, consisting of several simple machines, for the densification of fresh green weeds into a "green fuel" suitable for e.g. a household cooking stove.

This invention is a complete system for densification of green weeds with a comparably high water content into a solid fuel which can be stored and be used in e.g. household cooking stoves. The main components are:

- A weed chopper.
- An extruder.
- An open air solar dryer.

They are all simple, small and portable.

The raw material has been fresh land weeds of many kinds but also the roots from water hyacinths when the leaves have been used as cattle feed. The chopper has 2 (or 4) rotating cutter blades and is suitable for weeds with a moisture content of about 80% It is provided with a 2 Hp motor and can chop 150-180 kg of material per hour. The densifier is a form of extruder. It consists of a screw and a die and produces about 60 kg per hour of wet fuel in the form of round log-like pieces. It requires an electrical motor of 2 Hp. The solar drier consists of bottoms of sheet metal and covers of LDP plastic sheets and they function properly only on sunny days. It usually takes 3 days to reduce the moisture to 5% which is suitable for trouble-free storage and combustion.

This system upgrades waste to fuel but also offers an incentive for better weed control on farmland and in ponds.

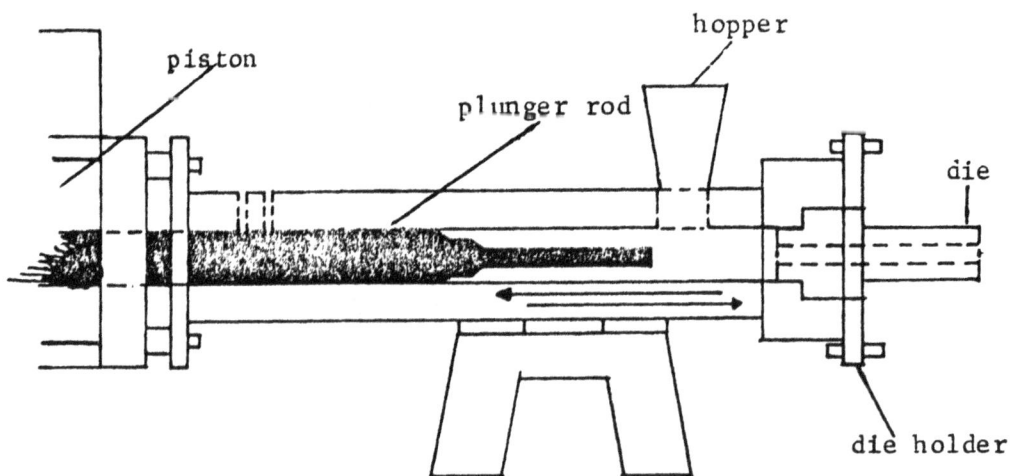

Contact: Mr Watna Stienswat, Department of Horticulture, Faculty of Agriculture, Kasetsart University, Bangkok, THAILAND.

NEW ENERGY VILLAGE
INTEGRATED BIOGAS PRODUCTION AND USE

Mr Huang Cong
Guangzhou Institute of
Energy Conversion
CHINA (PRC)

A communal system for the production and use of biogas and digester sludge in an agricultural village.

This innovation is a fully integrated system for the production and use of biogas in a Chinese agricultural village.

Several types of digesters (anaerobic fermenters) have been tested e.g. floating cover, elastomer cover and gasbag and also closed and pressurised from an overhead water tank.

The input consists of various waste materials of vegetable or animal origin, such as:

- Mulberry branches, sugar cane leaves, banana peels, mushroom waste etc.

- Pig manure, night soil, silk worm waste etc.

The output consists of biogas and digester sludge. The gas is used for heating but also drives a 12 kW electrical power plant. The sludge is fed to farmed fish together with napier grass but is also used for producing earthworms (chicken feed) and mushrooms.

A further advantage of the use of biogas digesters is that it prevents the recirculation of live parasitic organisms in the village.

Production Status in New Energy Village
Before/After Biogas Construction

Item \ Year	1977	1980		1981		1982		1983	
Population	442	448		446		460		458	
Per capita income (RMB¥)	240.5	444		409		627		809	
pig (kg)	346	523	+ 51%	637	+84.1%	920	+ 165.8%	773	+123.4%
Pond fish (kg)	69450	84500	+ 21.6%	83500	+ 20.2%	126000	+81.4%	166150	+139.2%
Sugarcane (kg)	55450	600300	+ 8.25%	550000	-0.81%	454050	-18%	325000	-41.4%
Mulberry leaf (kg)	180950	216900	+19.8%	198900	+ 9.9%	200000	+10.1%	153500	- 15.2%
Silkworm cocoon (kg)	10766	11850	+ 10.1%	13000	+ 20.1%	6600	-38.7%	6350	- 41%
Banana (kg)	14200	15300	+ 7.7%	25550	+79.9%	32250	+127%	35000	+146.5%

Contact: Mr Huang Cong, Guangzhou Institute of Energy Conversion, Chinese Academy of Sciences, Guangzhou, CHINA (PRC).

FIFTY INNOVATIONS IN SHORT

Space in this book does not allow us to give a full description of all the hundred best inventions nominated to IIA. Fifty of them therefore are limited to a descriptive title, the name(s) of the inventor(s) and an address to write to for further information.

WATER

THE ROWER PUMP

is a manually operated reciprocating pump.Mr. Georg Klassen, CANADA Contact: Mirpur Agricultural Workshop, Mirpur Section 12, Dacca, BANGLADESH. Telephone: 382544.

DIURNAL CYCLE SOLAR WATER PUMP

Contact: Mr. Roger Bernard, Université Lyon 1, 17 rue Laënnec, F-69300 Caluire, FRANCE. Telephone: +33-7-8898124 extension 3392

AIR EVACUATION UNIT FOR SIPHONS

in connection with a dam for irrigation and for small hydro electric power plants. Contact: Arne Fjälling, Institute of Freshwater Research, S-17011 Drottningholm, SWEDEN. Telephone: +46-8-7590040.

INDUSTRY

FARM SCALE FERTILISER PRODUCTION

by using surplus electricity from a mini-hydroelectric plant to produce nitrogen oxide. Contact: Mr Miguel Carlos Andreotti, Mr Paulo Evaristo da Silva Chaves and Mr Wilson Martins, Cidade Universitária Armando de Sales Oliveira, Butantã, 05508-São Paulo, BRAZIL.

PRODUCTION OF FRUIT IN THE TROPICS

Methods for regulation of the time of fruiting of trees in the tropics and the subtropics. Contact: Mr Jau-Chang, Institut für Obstbau und Baumschule, Universität Hannover, Am Steinberg 3, D-3203 Sarstedt, FRG.

ANIMAL FEED

from agricultural wastes, which are treated with urea and sodium hydroxide. Contact: Mr Do-Jeon Cho, 72 Inbong-dong, Sangju-up, Angju-gun, Kyongsangbuk-do, Seoul, SOUTH KOREA.

FARM IMPLEMENT FOR IMPROVED MANAGEMENT OF VERTISOLS

a type of African soil, which is difficult to farm. Contact: Dr. Samuel Jutzi or Mr Abiye Astatke, International Livestock Centre for Africa (ILCA), P.O.Box 5689, Addis Ababa, ETHIOPIA. Telephone: 183215.

MULTI-CROP THRESHER AND WINNOWER

An adaptation of the Asian Axial Flow Thresher to wheat and other crops. Contact: Mr Ahmed A. Bahgat, Catholic Relief Services, USCC, Egypt Program, P.O.Box 2410, Cairo, EGYPT.

MULTIPURPOSE SMALL FARM IMPLEMENT

provided with a four Hp engine. Mr Deoclécio Silveira Barros. Contact: Ms Dalva Lúcia Nobre, Director, Serviço Estadual de Assistência aos Inventores (SEDAI), Av.Angelica 2632, 9 andar, CEP-01228, São Paulo, BRAZIL. Telephone: +55-11-2555713

THE SINGLE-OX PLOUGH

An invention for the small Ethiopian farmer. Contact: Mr. Abiye Astatke, International Livestock Centre for Africa (ILCA), P.O.Box 5689, Addis Ababa, ETHIOPIA. Telephone: 183215.

CANE SUGAR EXTRACTOR

consisting of a continuous screw press. Antonio Cantizani Filho. Contact: Ms Dalva Lúcia Nobre, Director, Serviço Estadual de Assistência aos Inventores (SEDAI), Av. Angelica 2632, 9 andar, CEP–01228, São Paulo, BRAZIL. Telephone: +55-11-2555713

THE WADHWA PALM OIL PROCESSING PLANT

for the extraction of oil from palm fruits and nuts. Dr Narain Das Wadhwa. Contact: Mr E. Abrahams, AGRICO, P.O.Box 12127, Accra North, GHANA. Tel: 28260

DECENTRALIZED COTTON SPINNING

by means of simple small-scale machinery. Contact: Pastor I.J. Abraham, Peoples Church, Kazipet, Sharon Laxmipur, Warangal 506 013, INDIA.

AGENT FOR ACCELERATION OF CEMENT HARDENING

in tropical, hot and humid countries. Nguyen Minh Ngoc and Pham Van Trinh. Contact: Dr. An Khang, Director General, National Office on Inventions of the SR Vietnam, P.O.Box 432, 39 Tran Hung Dao Str., Hanoi, VIETNAM.

HURRICANE TILES

made locally of vegetable fibre concrete. A system developed by Mr. John P.M. Parry, Mr. Andrew Carrier and Mr. Tudus Wyn Jenkins. Contact: J.P.M. Parry & Associates Ltd. / IT Workshop, Overend Road, Cradley Heath, West Midlands B64 7DD, UK.

BUILDING ELEMENTS OF MUD-CONCRETE

A modular system for making building elements of cement and sandy soil. Dionísio Aurélio Caribé de Azevedo. Contact: Ms Dalva Lúcia Nobre, Director, Serviço Estadual de Assisténcia aos Inventores (SEDAI), Av.Angelica 2632, 9 andar, CEP-01228 São Paulo, BRAZIL. Telephone: +55-11-2555713

JOINING TECHNIQUE FOR RIGID CONSTRUCTION ELEMENTS

consisting of a system of grooves, wedges and springs. Contact: Mr. Rolf Schaefer, ROSCH System GmbH, Bauerstrasse 35, D-4100 Duisburg 1, FRG. Telephone: +49-203-333174.

FLOORING AND ROOFING SYSTEM

in ceramic material. Contact: Mr Altaf Hossain, Mirpur Ceramic Works Ltd., 20 Banga Bandhu Avenue, P.O.Box 19, Dhaka-2, BANGLADESH.

LIGHTWEIGHT HEAT STORING BUILDING SYSTEM

using local silicate based materials of construction. Dr Michael Párkányi. Contact: Dr István Bódis, Novotrade RT, Pf. 139, H-1389 Budapest, HUNGARY. Telephone: 530022. Patent granted in Hungary.

STORMPROOF HOUSES

built in groups of three for fishermen. Mr Vu Van Tan and Mr Hoang Huu Fe. Contact: Union of Architecture of Vietnam, 23 Dinh Tien Hoang, Hanoi, VIETNAM, with a sopy to Dr. An Khang, Director General, The National Office on Inventions of the SR Vietnam, P.O.Box 432, 39 Tran Hung Dao Str., Hanoi, VIETNAM.

BUILDING SYSTEM 5

requires a minimum of machinery and skilled labour in a team for construction. Contact: Dr Jawad Anani, President, Royal Scientific Society, P.O.Box 925819, Amman, JORDAN. Patent granted in Jordan.

POLES FOR AERIAL WIRES

such as telephone, telegraph or electricity. Made of phenol plastic in order to last in tropical climates. Contact: Mr Heinrich Pichler, Nachtigallenstrasse 7, A-5023 Salzburg, AUSTRIA. Telephone: +43-662-768082.

HOMOGENEOUS MICRO-AEROSOL

for eradication of insect pests and general sanitation. Contact: Mr Måns Arborelius, Viasol Marketing, Kastanjeallén 18, S-230 44 Bunkeflostrand, SWEDEN, Telephone: +46-40-153515.

FORESTRY

REFORESTATION OF MOVING SAND DUNES

in arid areas. Mr M.S. Hajjej (Tunisia) and Mr Axel Jensen (Denmark), Contact: Représentant Résident, PNUD, B.P. 620, Nouakchott, MAURITANIA.

NATURAL FOREST MANAGEMENT FOR SUSTAINED YIELD

Contact: Professor Dr. Roberto Tuyoshi Hosokawa, Department of Forest Management and Economics, Federal University of Paraná, P.O.Box 2197, 80 000 Curitiba, Paraná, BRAZIL. Telephone: +55-41-2331620

ALTERNATE CYCLE AGROFORESTRY ON MARGINAL HILL LANDS

A swidden system for extensive agriculture in sparsely populated areas with three-year annual food crops alternating with perennials for ten or more years. Contact: Dr Joseph A. Weinstock, 59 Palmear Road, Freeville, NY 13068, USA

ALCHI

A system for reforestation of arid areas based on a plough with two shares, which make it possible to plant two rows simultaneously. Contact: Dr Alfonso Alegría Jimenez, ICONA, calle 18 de Julio, E-16001 Cuenca, SPAIN.

VEGETATIVE PROPAGATION OF PINES FROM NEEDLE FASCICLES

A simple and easy method of cloning valuable individual forest trees. Contact: Dr. Xu Zhaoxiang, Deputy Director, NRCSTD, CHINA. (PRC)

THE GERRYCART AUTOMATIC POT CARRIER

invented by Mr. Gerardo M. Beltran and used in reforestation projects in Asia. Contact: The Chief, Social Forest Division, Bureau of Forest Development, Visayas Ave., Diliman, Quezon City, PHILIPPINES.

INCREMENT MEASUREMENT OF STANDING TREES

in a research or control forest. Dr Walter Bitterlich and Mr Benno Hesske. Contact: Dr Walter Bitterlich, Rennbahnstrasse 4 a, A-5020 Salzburg, AUSTRIA. Tel: +43-662-247654.

ENERGY

HOT AIR GENERATOR FOR DRYING CEREALS

fuelled with organic agricultural waste. Dr Atuo Sato and Mr Artur Zakis. Contact: Dr Atuo Sato, Instituto de Atividades Espaciais, 12.100 São José dos Campos, São Paulo, BRAZIL.

BIOGAS FROM MEAT PROCESSING

An integrated livestock and meat processing plant with extensive energy recovery. Contact: Mr Felix D. Maramba Sr., Liberty Bldg., Pasay Road, Makati, Metro Manila, PHILIPPINES. Telephone: 865011 or 885090

LINEAR FRESNEL LENS

and two more inventions relating to solar energy. Developed by Mr M. Maly and Mr B. Nábêlek. Patent granted in CZ. Contact: Institute of Physics, Czechoslovak Academy of Sciences, Na Slovance 2, Praha 8, CZECHOSLOVAKIA.

SUN RADIATION CONCENTRATOR WITH FLUID FRESNEL LENSES

which can be constructed locally. Contact: Mr. Thomas Weilenmann, Trädgårdsvägen 36, S-43361 Partille, SWEDEN. Telephone: +46-31-265050.

SOLAR WATER HEATER FOR WINDOWS IN THE TROPICS

Contact: Dr. J. T. Pytlinski, Senior Scientist, Center for Energy and Environment Research, University of Puerto Rico, G.P.O. Box 3682, San Juan, PUERTO RICO 00936, USA. Telephone: +1-809-7670350

ROTATING PRISMS SOLAR WALL

consisting of heavy standing triangular columns which are rotated in a 24-hour cycle. Contact: Professor David Faiman, Ben-Gurion University, Institute for Desert Research, Sede Boqer Campus, 84990 ISRAEL. Telephone: +972-57-35333.

SOLAR-POWERED VACCINE REFRIGERATOR

named BP VR50 and developed by Mr. C.C. Proctor, Mr. R.A. Brice and Mr. B. Light of British Petroleum. Contact: Miss L. Munnoch, BP Solar Systems Limited, Farmbrough Close, Stocklake, Aylesbury, Bucks HP20 1DQ, UK. Telephone: +44-296-26100.

MEDICAL VAN POWERED BY SOLAR AND WIND ENERGY

designed primarily for vaccination programmes in developing countries by Dr. P.G. Virapin and Dr. A. Monod. Contact: Service de Recherche Appliquée de T.M.S.S., 15 avenue De Maurin, F-34000 Montpellier, FRANCE. Telephone: +33-67-926901

ELECTRO-MECHANICAL FILM

A gas-filled biaxially orientated film coated with electrically conducting layers. A low-cost alternative to solar cells. Contact: Mr. Johannes Kirjavainen, Kristianink. 7C 38, Helsinki 17, FINLAND. Telephone: +358-0-639847

DIESEL ENGINE FUEL FROM COCONUT OIL

Patent granted in the Philippines. Contact: Mrs. Violeta P. Arida, National Institute of Science and Technology, Chemical Research and Development Center, Pedro Gil St., Ermita, P.O.Box 774, Manila, PHILIPPINES. Telephone: 503041.

PYROLYSIS OF RICE HUSKS

to produce fuel gas and charcoal briquettes. Contact: Mr. Nara Pitakarnnop, Director of Energy Technology Department, Thailand Institute of Scientific and Technological Research, 196 Phahonyothin Rd., Bangkhen, Bangkok 10900, THAILAND.

SINGLE PAN WOOD STOVE

A light, portable and efficient design developed by Mr. H.S. Mukunda and Mr. U. Shrinivasa. Contact: Department of Aerospace Engineering, Indian Institute of Science, Bangalore 560 012, INDIA. Telephone: 34411.

FAMILY COOKER / MEDICAL COOKER

developed by Dr. Ronald Gerrits, Mr. J.C. Overhaart and Mrs Fanny Rosenzweig. Contact: The Foundation for Ecological Development Alternatives, P.O.Box 168, NL-2040 AD Zandvoort, NETHERLANDS. Telephone: +31-2507-16296

COOKING STOVE FOR BEDOUIN ARABS

Contact: Mr. Claude Osborne Forestier-Walker, MINOLA Enterprises Ltd., Rosebank, 13 Park Road, Aldeburgh, Suffolk, UK. Telephone: Aldeburgh 2102.

COOKING STOVE OF MOULDED CERAMICS

The moulds are made centrally but the building blocks are cast locally. Contact: Mr. Lennart Eriksson, IDERIKSON HB, Floravägen 24, S-39359 Kalmar, SWEDEN. Telephone: +46-480-12700

COCONUT SHELL CARBONIZATION

for medium scale batch production of charcoal. Developed by Mr. G.R. Breag, Mr. A.P. Harker and Mr. A.R. Paddon and tested in Sri Lanka. Contact: The Director, Tropical Development & Research Institute, Culham, Abingdon, Oxon OX14 3DA, UK.

SOLAR HOT WATER STORAGE HEATER

Developed and produced in Burkina Faso. Contact: Centre Écologique Albert Schweitzer, Rue du Rocher 13, CH-2000 Neuchâtel, SWITZERLAND. Telephone: +41-38-250836.

STOVES FUELLED BY COARSE BIOMASS

A flower-pot stove and a brick-pile stove for burning rice hulls, fine leaves etc. Contact: Dr. Herman Johannes, Gadjah Mada University, Jalan Pandega Duta III/17, Yogyakarta, INDONESIA.

BIOMASS COOKING STOVE

with reduced fuel consumption and air pollution. Developed by Mr. Nazrul Islam of Bangladesh while working at Resource Systems Institute, East-West Center, Honolulu, Hawaii 96848, USA

BIFACIAL SOLAR CELLS AND LARGE APERTURE CONCENTRATOR

A simple design developed by Professor Luque at the Polytechnical University of Madrid. Contact: Isofoton SA, Polígono Industrial Santa Teresa, Caleta de Vélez 52, Málaga, SPAIN. Telephone: +34-952-340900.

INDEXES

It is our intention and hope that this book will be used as a source of information for people working in or for developing countries. In order to make it more practical as a reference book for appropriate technology we have included several indexes. Besides the subject index you will also find below an index of inventors, institutions and national origin.

SUBJECT INDEX

The Target Area 'Industry' has been split up into six smaller areas in order to facilitate the use of this index. A number of items have been placed in more than one area.

Building

Energy

Farming

Food

Forestry

Industry

Textiles

Transport

Water

Other Subjects

INVENTORS INDEX

INSTITUTION INDEX

NATIONALITY INDEX

www.ingramcontent.com/pod-product-compliance
Lightning Source LLC
Chambersburg PA
CBHW080628030426
42336CB00018B/3118